# Consensus of Multi-Agent Systems with Binary-Valued Measurements

王 婷 著

Beijing
Metallurgical Industry Press
2021

## Abstract

There are 6 chapters in this book. Chapter 1 introduces the identification of set-valued systems and consensus control of multi-agent systems. Chapter 2 introduces some mathematics knowledge needed for identification and consensus control. Chapter 3 introduces two kinds of identification methods for set-valued systems. Chapter 4 and Chapter 5 focus on the consensus problem with binary-valued observations under undirected topology and directed topology, respectively. Chapter 6 investigates the consensus problem of linear multi-agent systems with binary-valued observations.

The book is written for researchers and engineers working in control science and engineering, sensing network, and signal processing, etc.

**图书在版编目(CIP)数据**

二集值量测下的多智能体系统的同步控制＝Consensus of Multi-Agent Systems with Binary-Valued Measurements：英文/王婷著. —北京：冶金工业出版社，2021.3

ISBN 978-7-5024-8756-0

Ⅰ.①二… Ⅱ.①王… Ⅲ.①智能系统—同步控制系统—英文 Ⅳ.①TP273 ②TP18

中国版本图书馆 CIP 数据核字(2021)第 044524 号

出　版　人　苏长永
地　　　址　北京市东城区嵩祝院北巷 39 号　邮编 100009　电话 (010)64027926
网　　　址　www.cnmip.com.cn　电子信箱　yjcbs@cnmip.com.cn
责任编辑　戈　兰　美术编辑　彭子赫　版式设计　孙跃红
责任校对　石　静　责任印制　李玉山
ISBN 978-7-5024-8756-0
冶金工业出版社出版发行；各地新华书店经销；北京虎彩文化传播有限公司印刷
2021 年 3 月第 1 版，2021 年 3 月第 1 次印刷
169mm×239mm；9 印张；175 千字；136 页
56.00 元
冶金工业出版社　投稿电话　(010)64027932　投稿信箱　tougao@cnmip.com.cn
冶金工业出版社营销中心　电话　(010)64044283　传真　(010)64027893
冶金工业出版社天猫旗舰店　yjgycbs.tmall.com

(本书如有印装质量问题，本社营销中心负责退换)

# Preface

Over the last three decades, numerous research has been done on the cooperative control of multi-agent systems because of its wide applications. In a multi-agent system, all agents are required to cooperate with others to achieve a group objective and each of them can only get the local information from its neighbors. Considering account uncertainties, disturbances, modeling errors, etc, the local information is always set-valued in real world problems. Consequently, the agent can only obtain the set-valued information of their neighbors' states. The agent can only obtain the information whether the states of the neighbors belong to a set or some sets. How to use these local and set-valued information to achieve a group objective is a realistic and meaningful problem.

This book introduces how agents use local and set-valued information to achieve a group objective-consensus-by reaching an agreement on their states. Usually, these agents have topological distribution in realistic model. The undirected topology and directed topology are considered in this book. The multi-agent system is discrete. The first-order multi-agent system and the linear multi-agent system are introduced and discussed. Different consensus algorithms are desgined for different multi-agent systems. The main idea of the consensus algorithms is to design consensus control by the estimated states of neighbors. For the estimation, I introduce an offline identification algorithm and an online identification algorithm. Based on the offline and online estimation methods, I design two kinds of the consensus algorithms: two-time-scale consensus algorithm and recursive consensus algorithm. Finally, the convergence and convergence rate of the identification and control algorithms are analyzed.

My objective in writing this research book is to summarize my work in consensus control with binary-valued measurements. The book is written for researchers and engineers working in control science and engineering, sensing network, communication, signal processing, and mathematical statistic, etc. The book is mainly the author's understanding of consensus with binary-valued measurements. Interested authors can refer to some classic articles on quantized consensus in the references of the book.

The results in this book would not have been possible without the help of many people. In particular, I am indebted to Professor Yanlong Zhao at Academy of Mathematics and Systems Science, Chinese Academy of Sciences, for his leadership in the area of system identification with set-valued observations. I am also indebted to Professor Jin Guo at University of Science and Technology Beijing for many fruitful discussions on research ideas. I also acknowledge the efforts of Wenjian Bi, Ximei Wang, Hang Zhang, Min Hu, Ying Wang, Lifang Wu, Ruoshi Shi and Zhipeng Ren. I am thankful to my colleagues for providing me with an academic environment. Finally, I gratefully acknowledge the support of the National Natural Science Foundation of China under Grant 61803370.

Ting Wang<br>
Beijing, December 2020

# Notation

**R**: set of real numbers
**Z**: set of integers
**Z$^+$**: set of positive integers
**R$^n$**: Euclidean space of dimension $n$
**R$^{m\times n}$**: space of $m\times n$ real matrices
**$I_n$**: $n\times n$ identity matrix; also denoted as $I$ if the dimension is clear from the context
**$M^{-1}$**: inverse of $M$
**$M^T$**: transpose of $M$
$[A]_{ij}$: entry of matrix $A$ on $i$th row and $j$th column
$\mathrm{diag}(a_1, a_2, \cdots, a_n)$: diagonal matrix of $a_1, a_2, \cdots, a_n$
$E(x)$: mathematical expectation of random variable $x$
$\|x\|$: 2-norm of vector $x$; $\|x\| = (x^T x)^{1/2}$ unless indicated otherwise
$\|A\|_2$: the spectral norm of the matrix $A$
$\rho(A)$: the spectral radius of $A$
$\in$: belong to
$\otimes$: Kronecker product
$\prod_i x_i$: product of $x_i$
$\sum_i x_i$: sum of $x_i$
$|x|$: absolute value of $x$
$\lfloor x \rfloor$: floor function; the largest integer that is smaller or equal to $x$
$\lceil x \rceil$: ceil function; the least integer that is bigger or equal to $x$
$o(k)$: function of $k$ satisfying $\lim_{k\to 0}\dfrac{o(k)}{k}=0$
$O(\cdot)$: $f(k) = O(g(k))$ if there exists a positive number M and c such that $\left|\dfrac{f(k)}{g(k)}\right| \leq M$ for any $k>c$.
a. c.: almost certainly
i. i. d.: independent and identically distributed
w. p. 1: with probability one

# Contents

1 **Introduction** ........................................................ 1
   1.1  Motivation .................................................... 1
   1.2  Identification of Set-Valued Systems ................. 3
   1.3  Consensus Control of Multi-Agent Systems .......... 4
   1.4  Outline of the Book ........................................ 5

2 **Preliminaries** ...................................................... 6
   2.1  Vectors and Norms ......................................... 6
   2.2  Probability Theory ......................................... 8
   2.3  Algebraic Graph Theory ................................... 9
   2.4  Some Other Concepts ..................................... 10
   2.5  Notes ........................................................... 11

3 **Identification of Set-Valued Systems** ..................... 12
   3.1  Problem Formulation ...................................... 13
   3.2  Asymptotically Efficient Non-Truncated Identification Algorithm ...... 13
      3.2.1  Identification for Single-Parameter Systems ............ 14
      3.2.2  Identification for Multi-Parameter Systems ............ 24
      3.2.3  Numerical Simulation .................................... 29
   3.3  Recursive Projection Identification Algorithm ......... 32
      3.3.1  Algorithm Design ......................................... 32
      3.3.2  Properties of the Algorithm ............................ 33
      3.3.3  Numerical Simulation .................................... 42
   3.4  Notes ........................................................... 43

4 **Consensus with Binary-Valued Measurements under Undirected Topology** ........................................................ 44
   4.1  Problem Formulation ...................................... 45
   4.2  Two-Time-Scale Consensus ............................. 46

|  |  |  |
|---|---|---|
| 4.2.1 | Estimation | 46 |
| 4.2.2 | Consensus Control | 51 |
| 4.2.3 | The Consensus Protocol | 62 |
| 4.2.4 | Numerical Simulation | 63 |

4.3　Recursive Projection Consensus ········· 67

|  |  |  |
|---|---|---|
| 4.3.1 | Consensus Algorithm | 68 |
| 4.3.2 | Main Results | 72 |
| 4.3.3 | Numerical Simulation | 85 |

4.4　Notes ········· 89

## 5　Consensus with Binary-Valued Measurements under Directed Topology ········· 90

5.1　Problem Formulation ········· 90
5.2　Control Algorithm ········· 91
5.3　Properties of the Algorithm ········· 94

|  |  |  |
|---|---|---|
| 5.3.1 | Estimation | 94 |
| 5.3.2 | Transportation Design | 96 |

5.4　Numerical Simulation ········· 108
5.5　Notes ········· 112

## 6　Consensus of Linear Multi-Agent Systems with Binary-Valued Measurements ········· 113

6.1　Problem Formulation ········· 114
6.2　Review on the Case of Precise Communication ········· 115
6.3　Case of Binary-Valued Communication ········· 116

|  |  |  |
|---|---|---|
| 6.3.1 | Control Algorithm | 117 |
| 6.3.2 | Main Results | 119 |

6.4　Numerical Simulation ········· 129
6.5　Notes ········· 132

**References** ········· 133

# 1 Introduction

## 1.1 Motivation

Since the birth of cybernetics, the identification and control of uncertain systems have been the core issue of control science. Many internationally celebrated control scholars are devoted to the study of the identification and control of uncertain systems, such as R. E. Kalman, a member of the American Academy of Sciences, K. J. Astrom, a member of the Royal Swedish Academy of Sciences, and G. C. Goodwin, a member of the Australian Academy of Sciences. Over the past few decades, many mature theories and methods for typical system models and typical uncertainties have been developed, such as random filtering and random control theory for diffusion dynamics and Gaussian noise, robust control theory for bounded uncertainties, and parameter identification for unbounded uncertainties, etc. These results have become an important part of modern control theory.

However, a new type of uncertainty data—set-valued measurement data emerges in reality. Although it has wide applications, the attention it deserves has not been paid. A characteristic of this type of data is that it is impossible to obtain the precise value of the measured object, but only know whether the measured value belongs to a certain set. For examples, the oxygen sensor can only get whether the oxygen content is higher than a certain threshold when it measures automobile exhaust in the industry[1]; Schizophrenia data in biomedicine can only show whether an individual is "disease" or "healthy"[2]; Radar target recognition is concerned with whether the measured object is a "missile" or "bait"[3].

The systems in the above examples can be described as the set-valued system shown in Fig. 1.1, where the output of the system is the set-valued data. From the structural point of view, the biggest difference of the set-valued system from the traditional system is the set-valued sensor part. The set-valued sensor can be an oxygen sensor in the actual industry, or a virtual quantizer, or a comparator in biology. The set-valued sensor prevents us from obtaining the precise value of the system output, such as $y = 80$. We can only get some rough set-valued data, such as $y > 60$ or $y \leqslant 60$.

The multi-agent system is a kind of complex dynamic system that widely exists in na-

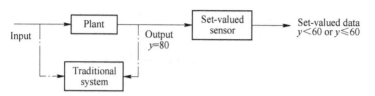

Fig. 1.1 The set-valued system

ture, engineering and social environment. It is an important mathematical model for exploring group phenomena such as self-organizing collective behavior of biological groups and social networks established by interpersonal relationships. The problem of consensus is one of the most basic and important distributed coordinated control of the multi-agent system, and it is also the basis of many distributed control and parameter estimation problems[4,5]. The theory of consensus has broad application prospects in civil, homeland security, and military fields. It can be directly applied to the coordinated control of multi-vehicles, including autonomous formation and formation maintenance of unmanned air, ground, and underground vehicles, target recognition and cooperative timing, etc[6~8].

One of the main characters of multi-agent systems is that the information is local, that is to say, each agent can only get the information from its neighbors, not the global information from all agents in the system. However, each agent often cannot get accurate information from its neighbors in actual systems due to the limitations of measurement capabilities or network transmission conditions. It can only get whether the information belongs to a certain set or some sets. If the agent in the system can only get whether the information from its neighbor belongs to one set, the system is called the multi-agent system with binary-valued measurements. The information available in multi-agent systems with binary-valued measurements is much less than that of the traditional multi-agent system with precise measurements. The most existing literatures design consensus control of multi-agent systems by using the accurate state of neighbors. Therefore, it is necessary and difficult to discuss how to use binary-valued measurements and local information to realize consensus control.

One natural idea is to design consensus control by the estimates of the states of neighbors since the accurate states are unknown with binary-valued measurements. So, each agent needs to estimate the states of its neighbors by using binary-valued information. The identification of set-valued systems may provide some estimation methods. Therefore, we will review the identification of set-valued systems and the consensus control of multi-agent systems in the following.

## 1.2 Identification of Set-Valued Systems

With wide driving forces, some key scientific problems on set-valued signals emerge such as how to model and control the system, how to generate a decision of true or false. As one of the bases to solve these problems, parameter identification based on set-valued signals has been paid a lot of attentions. Existing works on identification with set-valued signals can be classified to be three catalogs: One-time completed estimation algorithms, iterative estimation methods and recursive estimation methods.

For one-time completed estimation algorithms, empirical measure method is one of the earliest and most popular methods to identify systems with set-valued signals. It is the maximum likelihood estimation in the case that the inputs of the set-valued systems are periodic. It was proposed in [1] to investigate the identification errors, time complexity, input design, and impact of disturbances and unmodeled dynamics on identification accuracy and complexity for linear systems with binary-valued information. The methodology has been extended to the identification of Winer systems and Hammer systems[9,10]. Asymptotic efficiency of this method was studied in [11], where the inverse of the distributed function was required to be uniformly bounded.

In terms of set-valued data with general inputs, the main challenge is that the maximum likelihood function cannot be solved directly. Iterative parameter estimation algorithms provide good ways to give the numerical solution of the maximum likelihood function. Ref. [12] introduced the expectation maximization (EM) algorithm to estimate the parameter with quantized data and simulation results showed the convergence properties of the algorithm. Ref. [13] constructed an EM-type algorithm for finite impulse response systems with binary-valued outputs. More importantly, the algorithm is proved to be convergent with an exponential rate. This method has been applied to both complex diseases diagnosis to find the causal genes, and radar target recognition to separate the real targets from the baits.

The recursive ones are updated by the newest data which can save more calculations. Ref. [14] gave a recursive estimation method based on binary observations, the convergence were shown by simulated examples. Under the Gaussian assumption on the predicted density, Ref. [15] investigated the minimum mean square filtering using the set-valued innovations, where the threshold was design as the prediction of system outputs. Ref. [16] proposed a recursive estimation algorithm with low-storage requirements and computational complexity provided that the input signal satisfies a strong mixing property. Ref. [17] proposed a recursive projection algorithm for multiple parameters estimation with binary signals without periodic input constraints and the algorithm was proved to be convergent with a convergence rate O $(\log k/k)$.

## 1.3 Consensus Control of Multi-Agent Systems

Consensus control of multi-agent systems has attracted a lot of attentions recently due to its wide applications. At the beginning of consensus study, researchers mainly devoted themselves to dealing with local information with assumption of precise communication. They used the average rule to design a control to achieve consensus with connected topology.

However, measurement noises and quantization are unavoidable in the real communication network. To deal with random measurement noises, references [18 ~ 21] designed different consensus algorithms under different scenarios. Discrete – time and continuous-time consensus update schemes were proposed based on the discrete-time and continuous-time Kalman filters in [18]. Sufficient conditions were given for consensus seeking by using the proposed schemes. A stochastic approximation-type algorithm was introduced with the key feature of a decreasing step size in [19]. It was proved that the algorithm is convergent. Networks with random link failures and measurement noises were considered in [20]. A decreasing consensus gain as [19] is designed to attenuate the noises. In [21], a time-varying consensus gain is designed for the networks with continuous-time integrator agents.

To deal with consensus problems with different quantized communications, references [22~28] designed different consensus algorithms. Consensus problems with integer-valued states were firstly considered in references [22, 23]. They proposed algorithms to ensure the asymptotic convergence of agents' states to an integer approximation of the average of the initial states. Consensus problems with real-valued states and infinite-level quantized communications were considered in [24, 25]. A uniform quantization was considered in [24] and the properties of this algorithm were investigated both by a worst case analysis and by a probabilistic analysis. Ref. [25] proposed continuous-time and sampled-data-based protocols with the logarithmic quantization in the communications. Consensus problems with finite-level quantizers were studied in [26~28]. They assumed the communication channels were noiseless. Ref. [26] considered the truncated logarithmic quantizers with finite quantization levels. A novel approach was proposed and it was proved that the convergence factors depended on the quantization levels. In [27, 28], the distributed coordination problems of first-order systems and second-order systems were considered respectively. Especially in [27], an exponential convergence rate can be achieved by using 3 quantization levels.

From the reviews on quantized consensus, the existing literatures can deal with the consensus problems with at least 3-level quantization under noiseless communication. However, the case of binary-valued quantizer which is widely used in our daily life, such as oxygen sensors measuring the exhaust gas of an automobile[1], schizophrenia data in biomedicine[29], etc, cannot be solved by their methods. Besides, measurement noises are unavoidable in the real communication network. Therefore, how to achieve consensus under noisy and binary-valued communication is a more realistic and significant problem, which is studied in this book.

## 1.4　Outline of the Book

The book consists of five parts. The first part gives the motivation of the book and overviews identification and consensus control with set-valued measurements (Chapter 1). The second part introduces some preliminaries (Chapter 2). The third part introduces the identification of set-valued systems and two kinds of identification algorithms are given (Chapter 3). The fourth part introduces the consensus of first-order multi-agent systems, including the case of undirected topology (Chapter 4) and the case of directed topology (Chapter 5). The fifth part focuses on the consensus of linear multi-agent systems with binary-valued measurements (Chapter 6).

# 2 Preliminaries

## 2.1 Vectors and Norms

**Definition 2.1** On an $n$-dimensional space $R^n$, the Euclidean norm of the vector $x = (x_1, x_2, \cdots, x_n)^T$ is defined as the Euclidean distance from $x$ to zero, that is

$$\|x\| = \sqrt{x_1^2 + x_2^2 + \cdots + x_n^2}$$

In the book, the norms of vectors all mean the Euclidean norm.

A matrix norm on the vector space $R^{n \times n}$ is a function $R^{n \times n} \to R$ that must satisfy the following properties:

For all scalars $a \in R$ and for all matrices $A$ and $B$ in $R^{n \times n}$,

(1) $\|aA\| = |a| \|A\|$.
(2) $\|A+B\| \leq \|A\| + \|B\|$.
(3) $\|A\| \geq 0$, and $\|A\| = 0$ if and only if $A = 0$.
(4) $\|AB\| \leq \|A\| \|B\|$.

**Definition 2.2** The spectral norm of the matrix $A \in R^{m \times n}$ is defined as

$$\|A\|_2 = \sqrt{\lambda_{\max}(A^T A)}$$

The spectral norm of $A \in R^{m \times n}$ has the following properties:

(1) $\|A\|_2 = \max\limits_{\|x\|_2 = \|y\|_2 = 1} |y^T A x|$, $x \in R^n$, $y \in R^m$.
(2) $\|A^T\|_2 = \|A\|_2$.
(3) $\|A^T A\|_2 = \|A\|_2^2$.

**Definition 2.3** Assume $A \in R^{m \times n}$, $\lambda_1, \lambda_2, \cdots, \lambda_n$ are the eigenvalues of $A$, the spectral radius of $A$ is defined as

$$\rho(A) = \max_i |\lambda_i|$$

**Theorem 2.1** For any non-singular matrix $A \in R^{n \times n}$, the spectral norm of $A$ satisfies

$$\|A\|_2 = \sqrt{\rho(A^H A)} = \sqrt{\rho(AA^H)}$$

If $A$ is a normal matrix, then

$$\rho(A) = \|A\|_2$$

**Theorem 2.2** Assume the matrix $A \in R^{n \times n}$ is symmetric and nonnegative definite, $\lambda_1, \lambda_2, \cdots, \lambda_n$ are the eigenvalues of $A$, then

$$\lambda_1 \|x\|^2 \leqslant x^T A x \leqslant \lambda_n \|x\|^2, \quad \forall x \in R^n$$

where $\|x\|$ is the Euclidean norm of $x$.

**Definition 2.4** Assume $A = (a_{ij}) \in R^{m \times n}$, $B = (b_{ij}) \in R^{p \times q}$, then

$$A \otimes B = \begin{bmatrix} a_{11}B & a_{12}B & \cdots & a_{1n}B \\ a_{21}B & a_{22}B & \cdots & a_{2n}B \\ \vdots & \vdots & & \vdots \\ a_{m1}B & a_{m2}B & \cdots & a_{mn}B \end{bmatrix} \in R^{mp \times nq}$$

is Kronecker product of $A$.

The Kronecker product of $A$ has following properties:
(1) $k(A \otimes B) = kA \otimes B = A \otimes kB$, $k \in C$;
(2) $(A+B) \otimes C = A \otimes C + B \otimes C$;
(3) $(A \otimes B) \otimes C = A \otimes (B \otimes C)$.

**Theorem 2.3** Assume $A = (a_{ij})_{m \times n}$, $B = (b_{ij})_{s \times r}$, $C = (c_{ij})_{n \times p}$, $D = (d_{ij})_{r \times l}$, then

$$(A \otimes B)(C \otimes D) = AC \otimes BD$$

**Proof**

$$(A \otimes B)(C \otimes D) = (a_{ij}B)(c_{ij}D)$$

$$= (\sum_{k=1}^{n} a_{ij} c_{kj} BD) = (AC)_{ij} BD$$

$$= AC \otimes BD$$

**Theorem 2.4** Assume $A = (a_{ij})_{m \times n}$, $B = (b_{ij})_{p \times q}$ then

$$(A \otimes B)^T = A^T \otimes B^T$$
$$(A \otimes B)^H = A^H \otimes B^H$$

**Proof**

$$(A \otimes B)^T = (a_{ij}B)^T = \begin{bmatrix} a_{11}B & \cdots & a_{1n}B \\ \vdots & & \vdots \\ a_{m1}B & \cdots & A_{mn}B \end{bmatrix}^T = \begin{bmatrix} a_{11}B^T & \cdots & a_{m1}B^T \\ \vdots & & \vdots \\ a_{1n}B^T & \cdots & a_{mn}B^T \end{bmatrix} = A^T \otimes B^T$$

**Theorem 2.5** The matrices $A \in R^{m \times m}$ and $B \in R^{n \times n}$ are invertible. Then, the matrix $A \otimes B$ is invertible, and

$$(A \otimes B)^{-1} = A^{-1} \otimes B^{-1}$$

**Proof**

$$(A \otimes B)(A^{-1} \otimes B^{-1}) = (AA^{-1}) \otimes (BB^{-1}) = I_m \otimes I_n = I_{mn}$$

Then

$$(A \otimes B)^{-1} = A^{-1} \otimes B^{-1}$$

## 2.2 Probability Theory

**Definition 2.5** Suppose the sample space of a random experiment is $S$, a random variable $X$ is a function mapping from the sample space $S$ to the real set $R$.

**Definition 2.6** Suppose the distribution law of the discrete random variable $X$ is

$$P\{X = x_k\} = p_k, \quad k = 1, 2, \cdots$$

If the series

$$\sum_{k=1}^{\infty} x_k p_k$$

is absolute convergent, the sum of the series $\sum_{k=1}^{\infty} x_k p_k$ is called the mathematical expectation of the random variable $X$, denoted as $E(X)$, that is

$$E(X) = \sum_{k=1}^{\infty} x_k p_k$$

**Definition 2.7** Let the probability density of a continuous random variable $X$ be $f(x)$. If the integral

$$\int_{-\infty}^{\infty} xf(x) \, dx$$

is absolute convergent, the value of integral $\int_{-\infty}^{\infty} xf(x) \, dx$ is called the mathematical expectation of random variable $X$, which is denoted as $E(X)$, that is

$$E(X) = \int_{-\infty}^{\infty} xf(x) \, dx$$

**Theorem 2.6** (Schwarze inequality) For any random variables $X$ and $Y$, the following assertion holds

$$|E(XY)| \leq E|XY| \leq \sqrt{EX^2} \sqrt{EY^2}$$

if $EX^2 \leq \infty$, $EY^2 \leq \infty$.

**Theorem 2.7** (Chebyshev's inequality)  Let the mathematical expectation of random variable $X$ be $E(X)=\mu$, and the variance $D(X)=\sigma^2$. Then, for any $\varepsilon>0$, the following assertion holds.

$$P\{|X-\mu|\geq \varepsilon\} \leq \frac{\sigma^2}{\varepsilon^2} \quad \text{or} \quad P\{|X-\mu|<\varepsilon\} \geq 1-\frac{\sigma^2}{\varepsilon^2}$$

## 2.3  Algebraic Graph Theory

In mathematics, a graph is a structure described in a set of objects, some of which are "related" in a certain sense. A directed graph of order $p$ is a pair $(\gamma, \varphi)$, where $\gamma \triangleq \{1, \cdots, p\}$ is a finite nonempty node set and $\varphi \subseteq \gamma \times \gamma$ is an edge set of ordered pairs of nodes, called edges. Let define $\mathcal{G} \triangleq (\gamma, \varphi)$. Agent $j$ can obtain information from agent $i$, so we define the edge $(i, j)$ in the edge set of a directed graph. But it is not necessarily vice versa.

For a directed graph $\mathcal{G}=(\gamma, \varphi)$, if edge $(i, j) \in \varphi$, then $i$ is called $j$'s neighbor. The set of agent $j$'s neighbors is denoted by $N_j=\{i \in \gamma \mid (i, j) \in \varphi\}$. For an undirected graph $\mathcal{G}=(\gamma, \varphi)$, if edge $(i, j) \in \varphi$, then $i$ is called $j$'s neighbor and $j$ is called $i$'s neighbor.

The adjacency matrix is a matrix that represents the neighbor relationship between vertices. The adjacency matrix $\mathcal{A}(\mathcal{G})$ is the symmetric $n \times n$ matrix encoding of the adjacency relationships in the graph $\mathcal{G}$. Let denote $\mathcal{A}=[a_{ij}] \in \mathbb{R}^{N \times N}$ as the adjacent matrix of $\mathcal{G}$, $a_{ij}=1$ if $j \in N_i$, and $a_{ij}=0$ otherwise. Then, assume that $a_{ii}=0$, $i \in \gamma$. Denote $deg_i^{in} = \sum_{j=1}^{N} a_{ij}$, $deg_i^{out} = \sum_{j=1}^{N} a_{ji}$ as the in-degree and out-degree of node $i$, and $\mathcal{D}=\text{diag}(deg_1^{in}, \cdots, deg_N^{in})$ as the degree matrix of $\mathcal{G}$.

**Definition 2.8**  The Laplacian matrix $\mathcal{L}$ of $\mathcal{G}$ is defined as $\mathcal{L}=\mathcal{D}-\mathcal{A}$.

A sequence of edges in a directed graph of the form $(i_1, i_2)$, $(i_2, i_3)$, $\cdots$ is a directed path. Define analogously an undirected path in undirected graph. In a directed graph, directed path that starts and ends at the same node is a cycle. A directed graph is strongly connected if there is a directed path from every node to every other node. An undirected graph is connected if there is an undirected path between every pair of distinct nodes.

**Theorem 2.8**  Assume $0=\lambda_1 \leq \lambda_2 \leq \cdots \leq \lambda_n$ are the eigenvalues of Laplacian matrix $\mathcal{L}$ of graph $\mathcal{G}$. Graph $\mathcal{G}$ is connected if and only if $\lambda_2>0$.

**Example**  Consider the graph given in Fig. 2.1. The djacency matrix $A$, degree matrix $D$ and Laplacian matrix $L$ are given respectively as follows.

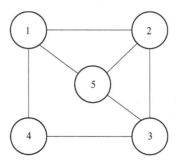

Fig. 2.1  A graph with 5 nodes

$$A = \begin{bmatrix} 0 & 1 & 0 & 1 & 1 \\ 1 & 0 & 1 & 0 & 1 \\ 0 & 1 & 0 & 1 & 1 \\ 1 & 0 & 1 & 0 & 0 \\ 1 & 1 & 1 & 0 & 0 \end{bmatrix},\ D = \begin{bmatrix} 3 & 0 & 0 & 0 & 0 \\ 0 & 3 & 0 & 0 & 0 \\ 0 & 0 & 3 & 0 & 0 \\ 0 & 0 & 0 & 2 & 0 \\ 0 & 0 & 0 & 0 & 3 \end{bmatrix},\ L = \begin{bmatrix} 3 & -1 & 0 & -1 & -1 \\ -1 & 3 & -1 & 0 & -1 \\ 0 & -1 & 3 & -1 & -1 \\ -1 & 0 & -1 & 2 & -1 \\ -1 & -1 & -1 & 0 & 3 \end{bmatrix}$$

## 2.4  Some Other Concepts

**Definition 2.9**  For sequences $f(k)$ and $g(k)$ with $k = 1, 2, \cdots$, $O(\cdot)$ is defined as

$$f(k) = O(g(k))$$

if there exists a positive number $M$ and $c$ such that $\left|\dfrac{f(k)}{g(k)}\right| \leq M$ for any $k > c$.

**Theorem 2.9**  [[30], Theorem 4.2.3]  If random variables $\{X_n \in L_p,\ n \geq 1\}$ satisfies

$$\sup_{m > n} E|X_m - X_n| = O(1), \text{ as } n \to \infty$$

Then, there exists a random variable $X$ such that

$$X_n \to X$$

in $L_p$ sense.

**Theorem 2.10**  For any given $\beta \in R$ and $r \in Z^+$, we have the following conclusions

(1)
$$\prod_{i=r}^{k}\left(1-\frac{\beta}{i}\right)=O\left(\frac{1}{k^{\beta}}\right),\quad k\to\infty$$

(2) For any $\delta>0$,

$$\sum_{l=r}^{k-1}\prod_{i=l+1}^{k}\left(1-\frac{\beta}{i}\right)\frac{1}{l^{1+\delta}}=\begin{cases}O\left(\dfrac{1}{k^{\delta}}\right)&\delta<\beta\\[4pt]O\left(\dfrac{\ln k}{k^{\beta}}\right)&\delta=\beta\\[4pt]O\left(\dfrac{1}{k^{\beta}}\right)&\delta>\beta\end{cases}$$

**Proof**

(1)
$$\prod_{i=r}^{k}\left(1-\frac{\beta}{i}\right)=\exp\left(\sum_{i=r}^{k}\log\left(1-\frac{\beta}{i}\right)\right)$$
$$=O\left(\exp\left(-\sum_{i=r}^{k}\frac{\beta}{i}\right)\right)$$
$$=O\left(\exp\left(-\beta\log\frac{k}{r}\right)\right)$$
$$=O\left(\frac{1}{k^{\beta}}\right)$$

(2)
$$\sum_{l=r}^{k-1}\prod_{i=l+1}^{k}\left(1-\frac{\beta}{i}\right)\frac{1}{l^{1+\delta}}=O\left(\sum_{l=r}^{k-1}\left(\frac{l}{k}\right)^{\beta}\frac{1}{l^{1+\delta}}\right)$$
$$=O\left(\frac{1}{k^{\beta}}\sum_{l=r}^{k-1}\frac{1}{l^{1+\delta-\beta}}\right)$$
$$=\begin{cases}O\left(\dfrac{1}{k^{\delta}}\right)&\delta<\beta\\[4pt]O\left(\dfrac{\ln k}{k^{\beta}}\right)&\delta=\beta\\[4pt]O\left(\dfrac{1}{k^{\beta}}\right)&\delta>\beta\end{cases}$$

## 2.5 Notes

The main content of Section 2.1 and theorems come from [31]. The main content of Section 2.2 comes from [32]. The main content of Section 2.3 comes from [33]. The interested reader is referred to them for detailed information.

# 3 Identification of Set-Valued Systems

Due to the limitation of communication resources, set-valued system has been highlighted in practical fields in recent years, and it has been widely used in industrial production, biopharmaceuticals, information industry and other fields[1].

The research on the identification of set-valued systems has been carried out for more than one decade, and many excellent works have been done, such as [1, 34, 35, 36]. Ref. [1] studies the identification of set-valued systems for the first time. It studies the effects of identification error, time complexity, input design and noise on identification accuracy of set-valued linear systems, and presents an identification algorithm based on empirical measure method. Refs. [11, 37] study the asymptotic effectiveness of this algorithm, but require that the inverse of the distribution function be uniformly bounded. In order to avoid this requirement, the empirical measure method of truncation is proposed in literature [10], and the asymptotic effectiveness of the algorithm is proved. However, how to give truncation depends on prior information.

In this chapter, we introduce two kinds of identification methods for set-valued systems: non-truncated identification algorithm and recursive projection identification algorithm. The non-truncated identification algorithm is an offline method. It is given by modifying the traditional empirical measure method[1]. The recursive projection identification algorithm[38] is an online method, which has faster speed and wider applications.

The non-truncated identification algorithm does not require the prior information. By using the discrete analysis property of binary measurement, we prove that the single parameter no-truncation identification is asymptotically unbiased and asymptotically effective. Furthermore, the multi-parameter identification is transformed into multiple single-parameter identification problems. Finally, the non-truncated identification algorithm is designed for multi-parameter systems, and the asymptotic effectiveness of the algorithm is proved.

The adaptive control usually requires that the input is not periodic, the empirical measure method and the non-truncated identification method are no longer applicable. Adaptive control requires an online identification algorithm, and Ref. [38] proposes a recursive projection identification algorithm. This chapter discusses a special form of the algorithm, that is, the form with step size of $1/k$, and proves that the algorithm has the convergence rate of $O(1/k)$, which is faster than that given in Ref. [38]. What's more, the convergence rate is proved to depend on the true parameter, instead of the prior information of the parameter in Ref. [38].

The structure of this chapter is as follows: Section 3.1 is the problem formulation. In Section 3.2, an asymptotically effective non-truncated identification algorithm is introduced. The convergence is analyzed and the convergence rate of the estimation error is given. Section 3.3 introduces the recursive projection identification algorithm, and the convergence and convergence rate are given, respectively. Section 3.4 is some notes.

## 3.1 Problem Formulation

Consider a linear system with binary-valued measurements:

$$\begin{cases} y(k) = \phi^T(k)\theta + d(k) \\ s(k) = I_{\{y(k) \leq C\}} \end{cases} \quad k = 1, 2, \cdots \quad (3.1)$$

where $\phi(k) \in R^n$ is the input, $\theta \in R^n$ is the unknown parameter, $d(k)$ is the noise, $C$ is a given threshold, $I_{\{\cdot\}}$ is an indicative function, specifically refers to

$$I_{\{y(k) \leq C\}} = \begin{cases} 1, & \text{if } y(k) \leq C \\ 0, & \text{else} \end{cases}$$

The purpose of this chapter is to estimate unknown parameters $\theta$ by using the input $\phi(k)$ and the output $s(k)$. Compared with the case of traditional accurate output with $y(k)$, the system with binary-valued output is nonlinear, and the information provided by each measurement is very limited.

**Assumption 3.1.** The noises $\{d(k), k = 1, \cdots, n\}$ are independent and identically normally distributed random variables with a known distribution function $F(\cdot)$, and the associated density function satisfies $f(x) = \mathrm{d}F(x)/\mathrm{d}x \neq 0$.

## 3.2 Asymptotically Efficient Non-Truncated Identification Algorithm

First, we introduce the definition of efficiency from estimation theory. Assuming $X_1$, $X_2, \cdots, X_N$ is a random sample of sample size $N$ from a random variable $X$ with probability density function $f(x; \nu)$, where $\nu$ is an unknown parameter. Let $\hat{\nu} = T(X_1, X_2, \cdots, X_N)$ be an unbiased estimator of $\nu$. Under certain regularity conditions (that are usually fulfilled for the identification problems we are working with), the variance $\sigma_{\hat{\nu}}^2$ of the estimator $\hat{\nu}$ is bounded by:

$$\sigma_{\hat{\nu}}^2 \geq \frac{1}{E\left[\dfrac{\partial \ln f(x_1, x_2, \cdots, x_N; \nu)}{\partial \nu}\right]^2}$$

The item $\dfrac{1}{E\left[\dfrac{\partial \ln f(x_1, x_2, \cdots, x_N; \nu)}{\partial \nu}\right]^2}$ is called the CR (Cramer-Rao) lower bound, denoted by $\sigma_{CR}^2$. If $X_1, X_2, \cdots, X_N$ are i.i.d. (independent and identically distributed), then the lower limit of CR can be written as:

$$\sigma_{CR}^2 = \frac{1}{NE\left[\dfrac{\partial \ln f(x; \nu)}{\partial \nu}\right]^2}$$

If $\sigma_{\hat{\nu}}^2 = \sigma_{CR}^2$, the estimator $\hat{\nu}$ is called efficient. The definition of asymptotically efficient is given in the limit sense.

**Definition 3.1** If the following two conditions are satisfied:

(1) The estimator $\hat{\nu}$ is asymptotically unbiased in sense that $E\hat{\nu} \to \nu$ as $N \to \infty$.

(2) $\dfrac{\sigma_{\hat{\nu}}^2}{\sigma_{CR}^2} \to 1$, as $N \to \infty$.

An estimator $\hat{\nu}$ with sample size $N$ is called asymptotically efficient.

For the case of multi-parameter (i.e. $\nu = [\nu_1, \cdots, \nu_N]^T$), the covariance matrix $\Sigma_{\hat{\nu}}$ of estimator $\hat{\nu} = T(X_1, X_2, \cdots, X_N)$ is bounded by

$$\Sigma_{\hat{\nu}} \geq \left[E\left(\frac{\partial \ln f(x_1, x_2, \cdots, x_N; \nu)}{\partial \nu}\right)\left(\frac{\partial \ln f(x_1, x_2, \cdots, x_N; \nu)}{\partial \nu}\right)^T\right]^{-1}$$

where $\left[E\left(\dfrac{\partial \ln f(x_1, x_2, \cdots, x_N; \nu)}{\partial \nu}\right)\left(\dfrac{\partial \ln f(x_1, x_2, \cdots, x_N; \nu)}{\partial \nu}\right)^T\right]^{-1}$ is the CR lower bound.

### 3.2.1 Identification for Single-Parameter Systems

Consider the system (3.1) with $y(k) = a_1 u(k-1) + d(k)$. Choose $u(k)$ to be a constant. Without loss of generality, assume $u(l) \equiv 1$. Then, system (3.1) can be written by

$$\begin{cases} y(k) = \theta + d(k) \\ s(k) = I_{\{y(k) \leq C\}} \end{cases} \quad k = 1, 2, \cdots \qquad (3.2)$$

where $\theta$ is an unknown parameter, $d(k)$ satisfies Assumption 3.1. How to estimate pa-

## 3.2 Asymptotically Efficient Non-Truncated Identification Algorithm

rameter $\theta$ by the binary-valued outputs $s(k)$, $k = 1, \cdots, N$?

For the identification of a single parameter system, we propose a non-truncation identification algorithm:

$$\psi_N = \frac{1}{N}\sum_{k=1}^{N} s(k); \quad \zeta_N = \begin{cases} \frac{1}{2} & \text{if } \psi_N = 0 \\ \psi_N & \text{if } 0 < \psi_N < 1; \quad \hat{\theta}_N^1 = C - F^{-1}(\zeta_N) \\ \frac{1}{2} & \text{if } \psi_N = 1 \end{cases} \quad (3.3)$$

where $\frac{1}{2}$ can be any other constants in the interval $(0, 1)$.

**Theorem 3.1** For single-parameter system (3.2), the estimate $\hat{\theta}_N^1$ given by Eq. (3.3) converges almost surely to the true value of the parameter $\theta$, that is

$$\hat{\theta}_N^1 \to \theta \quad \text{a. s.} \quad \text{as } N \to \infty$$

**Proof** Under Assumption 3.1, we have by Eq. (3.2)

$$E(s(k)) = P(y(k) \leq C) = P(d(k) \leq C - \theta) = F(C - \theta)$$

Denote $p = F(C - \theta)$. According to the strong law of large numbers, we have:

$$\psi_N = \frac{1}{N}\sum_{k=1}^{N} s(k) \to E(s(k)) = p \quad \text{a. s.} \quad \text{as } N \to \infty$$

By Assumption 3.1, the distribution function $F(\cdot)$ is strictly monotone and continuous. Thus, the inverse function $F^{-1}(\cdot)$ is continuous. Then, we have

$$\hat{\theta}_N^1 = C - F^{-1}(\zeta_N) \to C - F^{-1}(\delta) = \theta \quad \text{a. s.} \quad \text{as } N \to \infty$$

**Lemma 3.1** Given $N \in Z^+$, and $0 < p < 1$. $F(\cdot)$ is the distribution function of a Gaussian distribution. For any $\delta$ satisfied $0 < \delta < p < 1 - \delta < 1$, there exist positive constants $d_1$ and $d_2$ such that

$$Q_1 = \sum_{\frac{1}{N} \leq \frac{l}{N} < \delta} F^{-1}(l/N) C_N^l p^l (1-p)^{N-l} = O(e^{-d_1 N})$$

$$Q_2 = \sum_{1-\delta < \frac{l}{N} \leq \frac{N-1}{N}} F^{-1}(l/N) C_N^l p^l (1-p)^{N-l} = O(e^{-d_2 N})$$

**Proof** Firstly, let's see the $Q_1$

$$F^{-1}(l/N) C_N^l p^l (1-p)^{N-l}$$

$$= F^{-1}(l/N) \frac{N!}{l!(N-l)!} p^l (1-p)^{N-l}$$

Let
$$= \frac{N!}{l!(N-l)!} \delta^l (1-\delta)^{N-l} F^{-1}(l/N) \left(\frac{p}{\delta}\right)^l \left(\frac{1-p}{1-\delta}\right)^{N-l}$$

$$A = \frac{N!}{l!(N-l)!} \delta^l (1-\delta)^{N-l}, \quad B = F^{-1}(l/N) \left(\frac{p}{\delta}\right)^l \left(\frac{1-p}{1-\delta}\right)^{N-l}$$

Denote $g(v) \triangleq l \log v + (N-l) \log(1-v)$, then

$$g'(v) = \frac{l}{v} - \frac{N-l}{1-v} = \frac{N(l/N - v)}{v(1-v)}$$

Hence,

$$\log\left(\left(\frac{p}{\delta}\right)^l \left(\frac{1-p}{1-\delta}\right)^{N-l}\right)$$
$$= l \log p + (N-l) \log(1-p) - (l \log \delta + (N-l) \log(1-\delta))$$
$$= g(p) - g(\delta) = g'(v_1)(p - \delta), \quad \text{where } \delta < v_1 < p$$

Since $l/N < \delta$, it follows that

$$g'(v_1) = \frac{N(l/N - v_1)}{v_1(1-v_1)} < \frac{N(\delta - v_1)}{v_1(1-v_1)} \triangleq -\Delta_1 N$$

where $\Delta_1 = \frac{v_1 - \delta}{v_1(1-v_1)}$. Thus

$$\left(\frac{p}{\delta}\right)^l \left(\frac{1-p}{1-\delta}\right)^{N-l} \leq (e^{-\Delta_1(p-\delta)N}) = (e^{-\Delta_2 N}) \quad (3.4)$$

where $\Delta_2 = \Delta_1(p - \delta)$. Since $1/N \leq l/N < \delta$, we have

$$F^{-1}(1/N) \leq F^{-1}(l/N) < F^{-1}(\delta) \quad (3.5)$$

By Eq. (3.4) and Eq. (3.5), we have

$$|B| = |F^{-1}(l/N)| \left(\frac{p}{\delta}\right)^l \left(\frac{1-p}{1-\delta}\right)^{N-l} \leq \max\{|F^{-1}(1/N)| e^{-\Delta_2 N}, |F^{-1}(\delta)| e^{-\Delta_2 N}\}$$

On the other hand, we have

$$0 < A \leq \sum_{l=0}^{N} \frac{N!}{l!(N-l)!} \delta^l (1-\delta)^{N-l} = 1, \quad \text{for any } 0 \leq l \leq N \quad (3.6)$$

## 3.2 Asymptotically Efficient Non-Truncated Identification Algorithm

Thus

$$|Q_1| = \left| \sum_{1/N \leq l/N < \delta} AB \right| \leq \sum_{1/N \leq l/N < \delta} |B| \qquad (3.7)$$

$$\leq \max\{\delta N |F^{-1}(1/N)| e^{-\Delta_2 N}, \ \delta N |F^{-1}(\delta)| e^{-\Delta_2 N}\}$$

Let $F^{-1}(1/N) = w$ then $N = 1/F(w)$ Since $\lim_{w \to -\infty} F(w) = 0$, it follows that

Then
$$N \to \infty \Longleftrightarrow w \to -\infty$$

$$\lim_{N \to +\infty} N F^{-1}(1/N) e^{-\Delta_2 N} = \lim_{w \to -\infty} \frac{w}{F(w) e^{\Delta_2/F(w)}}$$

By L Hospital's Rule, we have

$$\lim_{w \to -\infty} \frac{w}{F(w) e^{\Delta_2/F(w)}}$$

$$= \lim_{w \to -\infty} \frac{1}{f(w) e^{\Delta_2/F(w)} - e^{\Delta_2/F(w)} \Delta_2 f(w)/F(w)}$$

$$= \lim_{w \to -\infty} \frac{F(w)}{(F(w) - \Delta_2) f(w) e^{\Delta_2/F(w)}} \qquad (3.8)$$

$$= \lim_{w \to -\infty} \frac{F(w)}{(F(w) - \Delta_2)} \lim_{w \to -\infty} e^{\frac{(w-a)^2}{2\sigma^2} - \frac{\Delta_2}{F(w)}}$$

$$= \lim_{w \to -\infty} e^{\frac{(w-a)^2 F(w) - 2\sigma^2 \Delta_2}{2\sigma^2 F(w)}}$$

Using L Hospital's Rule repeatedly, we have

$$\lim_{w \to -\infty} (w - a)^2 F(w) = \lim_{w \to -\infty} \frac{(w - a)^2}{1/F(w)}$$

$$= \lim_{w \to -\infty} \frac{-2(w - a) F^2(w)}{f(w)}$$

$$= \lim_{w \to -\infty} \frac{2F^2(w) + 4(w - a) f(w) F(w)}{(w - a) f(w)}$$

$$= \lim_{w \to -\infty} \frac{2F^2(w)}{(w - a) f(w)} = \lim_{w \to -\infty} \frac{4F(w) f(w)}{(1 - (w - a)^2) f(w)}$$

$$= \lim_{w \to -\infty} \frac{2f(w)}{-(w - a)} = \lim_{w \to -\infty} (w - a) f(w) = 0$$

Thus,

$$\lim_{w \to -\infty} (w - a)^2 F(w) - 2\sigma^2 \Delta_2 = -2\sigma^2 \Delta_2$$

together with $\lim_{w\to-\infty}\dfrac{1}{F(w)}=+\infty$ we have by Eq. (3.5)

$$\lim_{w\to-\infty}\frac{w}{F(w)e^{\Delta_2/F(w)}}=e^{-\infty}=0$$

Hence,

$$\lim_{N\to+\infty} NF^{-1}(1/N)e^{-\Delta_2 N}=\lim_{w\to-\infty}\frac{w}{F(w)e^{\Delta_2/F(w)}}=0$$

Analogously, we have

$$\lim_{N\to+\infty} NF^{-1}(1/N)e^{-\Delta_2 N/2}=0$$

which implies

$$N|F^{-1}(1/N)|e^{-\Delta_2 N}=O(e^{-\Delta_2 N/2}),\quad \text{as } N\to\infty$$

On the other hand, $Ne^{-\Delta_2 N}\to 0$ implies

$$N|F^{-1}(\delta)|e^{-\Delta_2 N}=O(e^{-\Delta_2 N/2}),\quad \text{as } N\to\infty$$

By Eq. (3.4), we have

$$Q_1=O(e^{-\Delta_2 N/2}) \qquad (3.9)$$

Now, let's see $Q_2$

$$Q_2=\sum_{1-\delta<\frac{l}{N}\leqslant\frac{N-1}{N}} F^{-1}(l/N)C_N^l p^l(1-p)^{N-l}$$

$$=\sum_{1-\delta<\frac{l}{N}\leqslant\frac{N-1}{N}} \frac{N!}{l!(N-l)!}(1-\delta)^l\delta^{N-l}F^{-1}(l/N)\left(\frac{p}{1-\delta}\right)^l\left(\frac{p}{\delta}\right)^{N-l}$$

Let

$$A'=\frac{N!}{l!(N-l)!}(1-\delta)^l\delta^{N-l}$$

$$B'=F^{-1}(l/N)\left(\frac{p}{1-\delta}\right)^l\left(\frac{1-p}{\delta}\right)^{N-l}$$

Analogously,

$$\log\left(\left(\frac{p}{1-\delta}\right)^l\left(\frac{1-p}{\delta}\right)^{N-l}\right)=g(p)-g(1-\delta)=g'(v_2)(p-(1-\delta))\leqslant -\Delta_3 N$$

where $v_2\in(p,1-\delta)$, $\Delta_3=\dfrac{(1-\delta-v_2)(1-\delta-p)}{v_2(1-v_2)}$.

$$\left(\frac{p}{1-\delta}\right)^l\left(\frac{1-p}{\delta}\right)^{N-l}\leqslant e^{-\Delta_3 N} \qquad (3.10)$$

Then

## 3.2 Asymptotically Efficient Non-Truncated Identification Algorithm

Since $(1-\delta)N < l \leq N-1$, we have

Thus
$$F^{-1}(1-\delta) < F^{-1}(l/N) \leq F^{-1}(1-1/N)$$

$$|B'| \leq \max\{|F^{-1}(1-\delta)|e^{-\Delta_3 N}, |F^{-1}(1-1/N)|e^{-\Delta_3 N}\}$$

Since

$$0 < A' < \sum_{l=0}^{N} \frac{N!}{l!(N-l)!}(1-\delta)^l \delta^{N-l} = 1 \tag{3.11}$$

We can obtain

$$|Q_2| \leq \sum_{1-\delta < \frac{l}{N} \leq \frac{N-1}{N}} |A'B'| \tag{3.12}$$

$$\leq \max\{\delta N |F^{-1}(1-\delta)|e^{-\Delta_3 N}, \delta N |F^{-1}(1-1/N)|e^{-\Delta_3 N}\}$$

Denote $w' = F^{-1}(1-1/N)$, then $N = \dfrac{1}{1-F(w')}$, we have

$$\lim_{N \to +\infty} F^{-1}(1-1/N) e^{-\Delta_3 N} = \lim_{w' \to +\infty} \frac{w'}{e^{\frac{\Delta_3}{1-F(w')}}} = \lim_{w' \to +\infty} \frac{1}{\dfrac{\Delta_3 f(w')}{(1-F(w'))^2} e^{\frac{\Delta_3}{1-F(w')}}} = 0$$

Hence

$$F^{-1}(1-1/N) e^{-\Delta_3 N} = O(e^{-(\Delta_3-1)N}) = O(e^{-\Delta_4 N})$$

where $\Delta_4 = \Delta_3 - 1$. Therefore,

$$\delta N F^{-1}(1-1/N) e^{-\Delta_3 N} = O(Ne^{-\Delta_4 N}) = O(e^{-\Delta_4 N/2}) \tag{3.13}$$

Together with

$$NF(1-\delta) e^{-\Delta_3 N} = O(e^{-\Delta_4 N})$$

We have by Eq. (3.9)

$$Q_2 = O(e^{-\Delta_4 N/2}) \tag{3.14}$$

Let $d_1 = \Delta_2/2$, $d_2 = \Delta_4/2$, we can get the lemma.

**Corollary 3.1** Given $N \in Z^+$, and $0 < p < 1$. $F(\cdot)$ is the distribution function of a Gaussian distribution. For any $\delta$ satisfying $0 < \delta < p < 1-\delta < 1$, there exist positive constants $d_3$ and $d_4$ such that

$$\sum_{\frac{1}{N} \leq \frac{l}{N} < \delta} F^{-1}(l/N)^2 C_N^l p^l (1-p)^{N-l} = O(e^{-d_3 N})$$

$$\sum_{1-\delta < \frac{l}{N} \leq \frac{N-1}{N}} F^{-1}(l/N)^2 C_N^l p^l (1-p)^{N-l} = O(e^{-d_4 N})$$

**Corollary 3.2** Given $N \in Z^+$, and $0 < p < 1$. For any $\delta$ satisfing $0 < \delta < p < 1 - \delta < 1$, there exist positive constants $d_5$ and $d_6$ such that

$$\sum_{\frac{1}{N} \leq \frac{l}{N} < \delta} C_N^l p^l (1-p)^{N-l} = O(e^{-d_5 N}), \quad \sum_{1-\delta < \frac{l}{N} \leq \frac{N-1}{N}} C_N^l p^l (1-p)^{N-l} = O(e^{-d_6 N})$$

Here are the main results.

**Theorem 3.2** (Asymptotic unbias) The estimator $\hat{\theta}_N^1$ given by Eq. (3.3) is asymptotically unbias in the sense that $E\hat{\theta}_N^1 \to \theta$, as $N \to \infty$. Moreover

$$E(\hat{\theta}_N^1 - \theta) = O\left(\frac{1}{N}\right)$$

**Proof** Denote $p = F(C - \theta)$ we have by Eq. (3.3)

$$E(\hat{\theta}_N^1 - \theta)$$
$$= E(C - \theta - F^{-1}(\zeta_N))$$
$$= \sum_{l=1}^{N-1} \left(C - \theta - F^{-1}\left(\frac{l}{N}\right)\right) C_N^l p^l (1-p)^{N-l} + \left(C - \theta - F^{-1}\left(\frac{1}{2}\right)\right)(p^N + (1-p)^N)$$
$$= \sum_{l=1}^{N-1} \left(C - \theta - F^{-1}\left(\frac{l}{N}\right)\right) C_N^l p^l (1-p)^{N-l} + O(e^{-d_7 N})$$

(3.15)

where $d_7 = \min\{\ln(1/p), \ln(1/(1-p))\}$. Dividing $\sum_{l=1}^{N-1} \left(C - \theta - F^{-1}\left(\frac{l}{N}\right)\right) C_N^l p^l (1-p)^{N-l}$ into three parts by $\delta$ ($0 < \delta < p < 1 - \delta < 1$), we obtain

$$\sum_{l=1}^{N-1} \left(C - \theta - F^{-1}\left(\frac{l}{N}\right)\right) C_N^l p^l (1-p)^{N-l}$$
$$= \left(\sum_{\frac{1}{N} \leq \frac{l}{N} < \delta} + \sum_{\delta \leq \frac{l}{N} \leq 1-\delta} + \sum_{1-\delta < \frac{l}{N} \leq \frac{N-1}{N}}\right)\left(F^{-1}\left(\frac{l}{N}\right) - (C - \theta)\right) C_N^l p^l (1-p)^{N-l}$$

By Lemma 3.1 and Corollary 3.2, we can get

$$\sum_{l=1}^{N-1} \left(C - \theta - F^{-1}\left(\frac{l}{N}\right)\right) C_N^l p^l (1-p)^{N-l}$$
$$= \sum_{\delta \leq \frac{l}{N} \leq 1-\delta} \left(C - \theta - F^{-1}\left(\frac{l}{N}\right)\right) C_N^l p^l (1-p)^{N-l} + O(e^{-d_1 N}) +$$
$$O(e^{-d_2 N}) + O(e^{-d_5 N}) + O(e^{-d_6 N})$$
$$= \sum_{\delta \leq \frac{l}{N} \leq 1-\delta} \left(C - \theta - F^{-1}\left(\frac{l}{N}\right)\right) C_N^l p^l (1-p)^{N-l} + O(e^{-d_8 N}) \quad (3.16)$$

## 3.2 Asymptotically Efficient Non-Truncated Identification Algorithm

where $d_8 = \min\{d_1, d_2, d_5, d_6\}$. Thus, we have by Eq. (3.4) and Eq. (3.5)

$$|E(\hat{\theta}_N^1 - \theta)| = \left|\sum_{\delta \leq \frac{l}{N} \leq 1-\delta} \left(C - \theta - F^{-1}\left(\frac{l}{N}\right)\right) C_N^l p^l (1-p)^{N-l}\right| + O(e^{-d_9 N})$$

(3.17)

where $d_9 = \min\{d_7, d_8\}$

Denote $G(\cdot) = F^{-1}(\cdot)$ By the well-known Taylor's formula, there exists a constant $\eta$ on the line segment joining $\frac{l}{N}$ and $p$ such that

$$F^{-1}\left(\frac{l}{N}\right) = F^{-1}(p) + G'(p)\left(\frac{l}{N} - p\right) + \frac{G''(\eta)}{2}\left(\frac{l}{N} - p\right)^2$$

$$= (C - \theta) + \frac{1}{f(C-\theta)}\left(\frac{l}{N} - p\right) + \frac{G''(\eta)}{2}\left(\frac{l}{N} - p\right)^2$$

(3.18)

Also, there exists a constant $M$ such that $|G''(\eta)| \leq M$ if $\delta \leq \frac{l}{N} \leq 1-\delta$. We have

$$\left|\sum_{\delta \leq \frac{l}{N} \leq 1-\delta} \left(C - \theta - F^{-1}\left(\frac{l}{N}\right)\right) C_N^l p^l (1-p)^{N-l}\right|$$

$$= \sum_{\varepsilon \leq \frac{l}{N} \leq 1} \left(\frac{1}{f(C-\theta)}\left(\frac{l}{N} - p\right) + \frac{G''(\eta)}{2}\left(\frac{l}{N} - p\right)^2\right) C_N^l p^l (1-p)^{N-l}$$

$$\leq \frac{1}{f(C-\theta)}\left|\sum_{\delta \leq \frac{l}{N} \leq 1-\delta}\left(\frac{l}{N} - p\right) C_N^l p^l (1-p)^{N-l}\right| +$$

$$\frac{M}{2}\left|\sum_{\delta \leq \frac{l}{N} \leq 1-\delta}\left(\frac{l}{N} - p\right)^2 C_N^l p^l (1-p)^{N-l}\right|$$

$$\leq \frac{1}{f(C-\theta)}\left|\left(\sum_{0 \leq \frac{l}{N} \leq 1} - \sum_{\frac{l}{N} < \delta} - \sum_{\frac{l}{N} > 1-\varepsilon}\right)\left(\frac{l}{N} - p\right) C_N^l p^l (1-p)^{N-l}\right| +$$

$$\frac{M}{2N^2}\left|\sum_{l=0}^{N}(l - Np)^2 C_N^l p^l (1-p)^{N-l}\right|$$

$$= \frac{1}{f(C-\theta)}\left|\sum_{0 \leq \frac{l}{N} < \delta}\left(\frac{l}{N} - p\right) C_N^l p^l (1-p)^{N-l} + \right.$$

$$\left.\sum_{1-\delta < \frac{l}{N} \leq 1}\left(\frac{l}{N} - p\right) C_N^l p^l (1-p)^{N-l}\right| + \frac{M\delta(1-\delta)}{2N}$$

$$= O(e^{d_5 N}) + O(e^{-d_6 N}) + O\left(\frac{1}{N}\right) = O\left(\frac{1}{N}\right)$$

By Eq. (3.17), we have

$$E(\hat{\theta}_N^1 - \theta) = O\left(\frac{1}{N}\right)$$

which implies $E\hat{\theta}_N^1 \to \theta$.

**Lemma 3.2** (CR lower bound) The CR lower bound for estimating $\theta$ based on observations $\{s_k,\ k = 1,\ \cdots,\ N\}$ in single-parameter system is

$$\sigma_{CR}^2(\theta,\ N) = \frac{F(C - \theta)(1 - F(C - \theta))}{Nf(C - \theta)^2}$$

**Proof** Let $s_k$ be some possible sample values of $s(k)$. Since $s(k)$ is a discrete random variable, the likelihood function of $s(k)$ taking value $s_k$, conditioned on $\theta$, is given by

$$l(s_k;\ \theta) = P(s_k;\ \theta) = F(C - \theta)^{s_k}(1 - F(C - \theta))^{1-s_k}$$

Under Assumption 3.1, we have $s(k)$, $k = 1,\ \cdots,\ N$ is independent identical distributed (i.i.d.). Then, the CR lower bound is given by

$$\sigma_{CR}^2 = \frac{1}{NE\left[\dfrac{\partial \ln P(s_k;\ \theta)}{\partial \theta}\right]^2} \tag{3.19}$$

Since

$$E\left[\frac{\partial \ln P(s_k;\ \theta)}{\partial \theta}\right]^2 = E\left[\frac{\partial(s_k \ln F(C - \theta) + (1 - s_k)\ln(1 - F(C - \theta)))}{\partial \theta}\right]^2$$

$$= E\left[-\frac{s_k f(C - \theta)}{F(C - \theta)} + \frac{(1 - s_k)f(C - \theta)}{1 - F(C - \theta)}\right]^2$$

$$= \frac{f(C - \theta)^2}{F(C - \theta)^2(1 - F(C - \theta))^2}E(s_k^2 - 2s_k F(C - \theta) + F(C - \theta)^2)$$

and

$$Es_k = F(C - \theta),\quad Es_k^2 = F(C - \theta)$$

We have

$$E\left[\frac{\partial \ln P(s_k;\ \theta)}{\partial \theta}\right]^2 = \frac{f(C - \theta)^2}{F(C - \theta)(1 - F(C - \theta))}$$

By Eq. (3.19), the CR lower bound is

## 3.2 Asymptotically Efficient Non-Truncated Identification Algorithm · 23 ·

$$\sigma_{CR}^2(\theta, N) = \frac{F(C-\theta)(1-F(C-\theta))}{Nf(C-\theta)^2}$$

**Theorem 3.3** (Asymptotic effectiveness) Under Assumption 3.1, the mean square error of estimator $\hat{\theta}_N^1$ given by Eq. (3.3) is

$$E(\hat{\theta}_N^1 - \theta)^2 = \frac{F(C-\theta)(1-F(C-\theta))}{Nf(C-\theta)^2} + O\left(\frac{1}{N^2}\right)$$

Moreover, estimator $\hat{\theta}_N^1$ is asymptotically efficient in sense that

$$N(E(\hat{\theta}_N^1 - \theta)^2 - \sigma_{CR}^2(\theta, N)) \to 0, \quad \text{as } N \to \infty$$

**Proof** Denote $p = F(C - \theta)$, we have by Eq. (3.3)

$$E(\hat{\theta}_N^1 - \theta)^2 = E(C - \theta - F^{-1}(\zeta_N))^2$$

$$= \sum_{l=1}^{N-1}\left(C - \theta - F^{-1}\left(\frac{l}{N}\right)\right)^2 C_N^l p^l (1-p)^{N-l} +$$

$$\left(C - \theta - F^{-1}\left(\frac{1}{2}\right)\right)^2 (p^N + (1-p)^N)$$

$$= \sum_{l=1}^{N-1}\left(C - \theta - F^{-1}\left(\frac{l}{N}\right)\right)^2 C_N^l p^l (1-p)^{N-l} + O(e^{-d_7 N})$$

where $d_7 = \min\{\ln(1/p), \ln(1/(1-p))\}$. Similar to the proof of Theorem 3.2, we divide the item $\sum_{l=1}^{N-1}\left(C - \theta - F^{-1}\left(\frac{l}{N}\right)\right)^2 C_N^l p^l (1-p)^{N-l}$ into three parts. According to Lemma 3.1, Corollaries 3.1 and 3.2, we can obtain

$$E(\hat{\theta}_N^1 - \theta)^2 = \sum_{\delta \leq \frac{l}{N} \leq 1-\delta}\left(C - \theta - F^{-1}\left(\frac{l}{N}\right)\right)^2 C_N^l p^l (1-p)^{N-l} + O(e^{-d_{10} N})$$

(3.20)

where $d_{10} = \min\{d_1, d_2, d_3, d_4, d_5, d_6, d_7\}$. By the Taylor's formula (3.18), We have

$$\sum_{\delta \leq \frac{l}{N} \leq 1-\delta}\left(C - \theta - F^{-1}\left(\frac{l}{N}\right)\right)^2 C_N^l p^l (1-p)^{N-l}$$

$$= \sum_{\delta \leq \frac{l}{N} \leq 1-\delta}\left(\frac{1}{f(C-\theta)}\left(\frac{l}{N} - p\right) + \frac{G''(\eta)}{2}\left(\frac{l}{N} - p\right)^2\right)^2 C_N^l p^l (1-p)^{N-l}$$

$$= \frac{1}{f(C-\theta)^2}\sum_{\delta \leq \frac{l}{N} \leq 1-\delta}\left(\frac{l}{N} - p\right)^2 C_N^l p^l (1-p)^{N-l} +$$

$$\sum_{\delta \leq \frac{l}{N} \leq 1-\delta} G''(\eta) f(C-\theta) \left(\frac{l}{N} - p\right)^3 C_N^l p^l (1-p)^{N-l} +$$

$$\frac{1}{4} \sum_{\delta \leq \frac{l}{N} \leq 1-\delta} G''(\eta)^2 \left(\frac{l}{N} - p\right)^4 C_N^l p^l (1-p)^{N-l}$$

$$\leq \frac{1}{f(C-\theta)^2} E\left(\frac{1}{N}\sum_{i=1}^{N} s_i - p\right)^2 + \frac{M}{f(C-\theta)} E\left|\frac{1}{N}\sum_{i=1}^{N} s_i - p\right|^3 +$$

$$\frac{M^2}{4} E\left(\frac{1}{N}\sum_{i=1}^{N} s_i - p\right)^4$$

$$= \frac{p(1-p)}{Nf(C-\theta)^2} + O\left(\frac{1}{N^2}\right)$$

where $\{s_i, i=1, \cdots, N\}$ is i.i.d., with $P(s_1 = 1) = p$, $P(s_1 = 0) = 1 - p$. Thus, we can get the mean square error as follows by Eq. (3.20)

$$E(\hat{\theta}_N - \theta)^2 = \frac{F(C-\theta)(1 - F(C-\theta))}{Nf(C-\theta)^2} + O\left(\frac{1}{N^2}\right)$$

By Lemma 3.2, we have

$$N(E(\hat{\theta}_N^l - \theta)^2 - \sigma_{CR}^2(\theta, N)) \to 0$$

which implies the estimator is asymptotically efficient.

### 3.2.2 Identification for Multi-Parameter Systems

Consider the multi-parameter system with binary-valued outputs:

$$\begin{cases} y(k) = \phi^T(k)\theta + d(k) \\ s(k) = I_{y(k) \leq C} \end{cases} \quad k = 1, 2, \cdots, \quad (3.21)$$

where $\phi(k) = [u(k-1), u(k-2), \cdots, u(k-n)]^T$ is the input, $\theta = [a_1, a_2, \cdots, a_n]^T$ is the unknown parameter, $n \geq 1$, How to estimate parameter $\theta$?

The main idea is transforming the multi-parameter identification problem to multiple single-parameter identification problems. Through estimating each component of the parameter $\theta$ independently, we can give the estimate of parameter $\theta$.

Since $\phi_k$ is defined by $\phi(k) = [u(k-1), u(k-2), \cdots, u(k-n)]^T$, we can choose $\phi(ln+1) = [1, 0, \cdots, 0]^T$, $\phi(ln+2) = [0, 1, \cdots, 0]^T$, $\cdots$, $\phi(ln+n) = [0, 0, \cdots, 0, 1]^T$, for $l = 1, \cdots, N$. At the skipping time $k = ln + 1$, $l = 1, \cdots, N$, the system (3.21) can be transformed into a single-parameter system:

## 3.2 Asymptotically Efficient Non-Truncated Identification Algorithm

$$\begin{cases} y(k) = a_1 + d(k) \\ s(k) = \mathbb{I}_{\{y(k) \leq C\}} \end{cases} \quad k = ln + 1, \; l = 1, 2, \cdots, N$$

By the identification algorithm in Section 2.3.1, we can get the estimator for $a_1$ as follows

$$\psi_N^1 = \frac{1}{N} \sum_{l=1}^{N} s(ln + 1), \quad \zeta_N^1 = \begin{cases} \frac{1}{2} & \text{if } \psi_N^1 = 0 \\ \psi_N^1 & \text{if } 0 < \psi_N^1 < 1, \; \hat{a}_1 = C - F^{-1}(\zeta_N^1) \\ \frac{1}{2} & \text{if } \psi_N^1 = 1 \end{cases}$$

Analogously, we can estimate the other parameter $a_i$, $i = 2, \cdots, n$ by

$$\psi_N^i = \frac{1}{N} \sum_{l=1}^{N} s(ln + i), \quad \zeta_N^i = \begin{cases} \frac{1}{2} & \text{if } \psi_N^i = 0 \\ \psi_N^i & \text{if } 0 < \psi_N^i < 1, \; \hat{a}_i = C - F^{-1}(\zeta_N^i) \\ \frac{1}{2} & \text{if } \psi_N^i = 1 \end{cases}$$

Then, the estimator for the parameter $\theta$ can be given by

$$\hat{\theta}_N^I = (\hat{a}_1, \cdots, \hat{a}_n)^T \tag{3.22}$$

The above estimator can be written in a vector form. Let

$$Y(l) = [y(ln + 1), y(ln + 2), \cdots, y((l+1)n)]^T$$
$$\Phi(l) = [\phi(ln + 1), \phi(ln + 2), \cdots, \phi((l+1)n)]^T$$
$$D(l) = [d(ln + 1), d(ln + 2), \cdots, d((l+1)n)]^T$$
$$S(l) = [s(ln + 1), s(ln + 2), \cdots, s((l+1)n)]^T$$

The system (3.21) can be written as

$$\begin{cases} Y(l) = \Phi(l)\theta + D(l) \\ S(l) = \mathbb{I}_{Y(l) \leq C} \end{cases} \quad l = 1, 2, \cdots, N \tag{3.23}$$

where $\mathbb{I}_{\{a \leq C\}} = [I_{\{a(1) \leq C\}}, I_{\{a(2) \leq C\}}, \cdots, I_{\{a(n) \leq C\}}]^T$ for any $n$-dimensional vector $a = [a(1), \cdots, a(n)]^T$. If we choose $\Phi(l) = I_{n \times n}$ for $l = 1, \cdots, N$, the estimator (3.11) can be written as follows.

$$\Psi_N = [\Psi_N(1), \cdots, \Psi_N(n)]^T = \frac{1}{N} \sum_{l=1}^{N} S(l)$$

$$\Xi_N = \Xi_N(i) = \begin{bmatrix} \frac{1}{2} & \text{if } \Psi_N(i) = 0 \\ \Psi_N(i) & \text{if } 0 < \Psi_N(i) < 1 \\ \frac{1}{2} & \text{if } \Psi_N(i) = 1 \end{bmatrix}_{n \times 1} \quad (3.24)$$

$$\hat{\theta}_N^I = C \mathbb{I}_n - F^{-1}(\Xi_N)$$

where $\mathbb{I}_n$ is the $n$-dimensional vector with all the elements being one, $F^{-1}(\Xi_N) = [F^{-1}(\Xi_N(1)), \cdots, F^{-1}(\Xi_N(n))]^T$.

The estimator $\hat{\theta}_N^I$ is given under the condition $\Phi(l) = I_{n \times n}$, which means the inputs $u(k)$, $k = 1, \cdots, (N+1)n$ are $n$-periodic with $\{u(1) = 0, \cdots, u(n-1) = 0, u(n) = 1\}$. If the inputs $u(k)$, $k = 1, \cdots, (N+1)n$ are $n$-periodic with other values, can we estimate parameter $\theta$? The answer is yes. We can give the estimator under the following assumption.

**Assumption 3.2** The inputs $u(k)$, $k = 1, \cdots, (n+1)N$ are $n$-periodic with $\{u(1) = u_1, \cdots, u(n) = u_n\}$. And $u = [u_n, u_{n-1}, \cdots, u_1]^T$ is full rank.

The definition of "full rank" for a vector is given as follows.

**Definition 3.2** A vector $v = [v_n, \cdots, v_1]^T$ is called full rank if the circulant matrix generated from $v$

$$T(v) = \begin{bmatrix} v_n & v_{n-1} & v_{n-2} & \cdots & v_2 & v_1 \\ v_1 & v_n & v_{n-1} & \cdots & v_3 & v_2 \\ v_2 & v_1 & v_n & \cdots & v_4 & v_3 \\ \vdots & \vdots & \vdots & & \vdots & \vdots \\ v_{n-1} & v_{n-2} & v_{n-3} & \cdots & v_1 & v_n \end{bmatrix}$$

is full rank.

Under Assumption 3.2, the input matrix $\Phi(l) = T(u) \triangleq \Phi$ is full rank. The estimator for $\theta$ is given by

$$\hat{\theta}_N = \Phi^{-1}(C \mathbb{I}_n - F^{-1}(\Xi_N)) \quad (3.25)$$

If $n = 1$, and $u(i) \equiv 1$, $i = 1, 2, \cdots$ system (3.23) is a single-parameter system and estimator $\hat{\theta}_N$ in (3.25) degenerates to estimator $\hat{\theta}_N$ in (3.3), which has been studied in Section 3.3.1. If $\Phi = I_n$ estimator $\hat{\theta}_N$ in (3.25) is $\hat{\theta}_N^I$ in (3.24). Estimator (3.25) is more general.

**Theorem 3.4** (Convergence) For system (3.21) with binary-valued output, the estimator $\hat{\theta}_N$ given by (3.25) converges almost surely to the true value of parameter $\theta$,

that is
$$\hat{\theta}_N \to \theta, \quad \text{a. s.}, \quad \text{as } N \to \infty$$

**Proof** Under Assumption 3.2, input matrix $\Phi(l) \equiv \Phi$. By (3.23) and Theorem 3.1, we have
$$C \mathbb{I}_n - F^{-1}(\Xi_N) \to \Phi\theta$$

Thus
$$\Phi^{-1}(C \mathbb{I}_n - F^{-1}(\Xi_N)) \to \theta$$

which implies the theorem holds.

**Lemma 3.3** (CR lower bound) Under Assumptions 3.1 and 3.2, the CR lower bound for estimating $\theta$ based on observations $\{s_k, k = n+1, \cdots, n(N+1)\}$ in system (3.21) is
$$\Sigma^2_{CR}(\theta, N) = \frac{1}{N}\Phi^{-1}\Lambda(\Phi^T)^{-1}$$

where $\Lambda = \text{diag}\left\{\dfrac{F(C - \phi_1^T\theta)(1 - F(C - \phi_1^T\theta))}{f(C - \phi_1^T\theta)^2}, \cdots, \dfrac{F(C - \phi_n^T\theta)(1 - F(C - \phi_n^T\theta))}{f(C - \phi_n^T\theta)^2}\right\}$,

$\phi_i = \phi(n+i)$, $i = 1, \cdots, n$

**Proof** Under Assumption 3.1, input vector $\phi(k)$ is $n$-periodic with
$$\phi(ln + i) = \phi(n + i) = \phi_i, \quad \text{for } l = 1, \cdots, N, \quad i = 1, \cdots, n$$

Then, input matrix $\phi(l)$ will be
$$\Phi = [\phi_1, \phi_2, \cdots, \phi_n]^T$$

Under Assumption 3.1, the observations $\{s_i, i = n+1, \cdots, n+nN\}$ are i.i.d., we have
$$P(s_{n+1}, s_{n+2}, \cdots, s_{n+nN}; \theta) = \prod_{k=n+1}^{n+nN} F(C - \phi^T(k)\theta)^{s_k}(1 - F(C - \phi^T(k)\theta))^{1-s_k}$$

and
$$\ln P(s_{n+1}, s_{n+2}, \cdots, s_{n+nN}; \theta)$$
$$= \sum_{k=n+1}^{n+nN}(s_k \ln F(C - \phi^T(k)\theta) + (1 - s_k)\ln(1 - F(C - \phi^T(k)\theta)))$$

Then,
$$\frac{\partial \ln P(s_{n+1}, s_{n+2}, \cdots, s_{n+nN}; \theta)}{\partial \theta} = \sum_{k=n+1}^{n+nN} \frac{\phi(k)f(C - \phi^T(k)\theta)(F(C - \phi^T(k)\theta) - s_k)}{F(C - \phi^T(k)\theta)(1 - F(C - \phi^T(k)\theta))}$$

Denote $f_k = f(C - \phi^T(k)\theta)$, $F_k = F(C - \phi^T(k)\theta)$. We have

$$E\left(\frac{\partial \ln P(s_{n+1}, s_{n+2}, \cdots, s_{n+nN}; \theta)}{\partial \theta}\right)\left(\frac{\partial \ln P(s_{n+1}, s_{n+2}, \cdots, s_{n+nN}; \theta)}{\partial \theta}\right)^T$$

$$= E\left(\sum_{k=n+1}^{n+nN} \frac{\phi(k) f_k (F_k - s_k)}{F_k(1 - F_k)}\right)\left(\sum_{k=n+1}^{n+nN} \frac{\phi(k) f_k (F_k - s_k)}{F_k(1 - F_k)}\right)^T$$

$$= \sum_{k=n+1}^{n+nN} \frac{\phi(k)\phi^T(k) f_k^2 E(F_k - s_k)^2}{F_k^2(1 - F_k)^2}$$

$$= \sum_{k=n+1}^{n+nN} \frac{\phi(k)\phi^T(k) f_k^2}{F_k(1 - F_k)} = N\sum_{i=1}^{n} \frac{\phi_i \phi_i^T f(C - \phi_i^T\theta)^2}{F(C - \phi_i^T\theta)(1 - F(C - \phi_i^T\theta))} = N\phi^T \Lambda^{-1} \Phi$$

where $\Lambda = \text{diag}\left\{\frac{F(C - \phi_1^T\theta)(1 - F(C - \phi_1^T\theta))}{f(C - \phi_1^T\theta)^2}, \cdots, \frac{F(C - \phi_n^T\theta)(1 - F(C - \phi_n^T\theta))}{f(C - \phi_n^T\theta)^2}\right\}.$

Therefore, the CR lower bound is

$$\Sigma_{CR}^2(\theta, N) = \frac{1}{N}\Phi^{-1}\Lambda(\Phi^T)^{-1}$$

**Theorem 3.5** (Asymptotic efficiency)   Under Assumptions 3.1 and 3.2, the estimator $\hat{\theta}_N$ given by (3.25) is asymptotically efficient in sense that

$$N(E(\hat{\theta}_N - \theta)(\hat{\theta}_N - \theta)^T - \Sigma_{CR}^2(\theta, N)) \to 0_{n \times n}, \quad \text{as } N \to \infty$$

**Proof**   Under Assumption 3.1, input matrix $\Phi(l)$ is a constant matrix:

$$\Phi(l) = \Phi = [\phi_1, \phi_2, \cdots, \phi_n]^T$$

System (3.23) will be

$$\begin{cases} Y(l) = [\phi_1^T\theta, \phi_2^T\theta, \cdots, \phi_n^T\theta]^T + D(l) \\ S(l) = \mathbb{I}_{\{Y(l) \leq C\}} \end{cases} \quad l = 1, 2, \cdots, N$$

By estimator (3.25), we have

$$\Phi\hat{\theta}_N = C\mathbb{I}_n - F^{-1}(\Xi_N)$$

where

$$\Xi_N = \Xi_N(i) = \begin{cases} \frac{1}{2} & \text{if } \Psi_N(i) = 0 \\ \Psi_N(i) & \text{if } 0 < \Psi_N(i) < 1 \\ \frac{1}{2} & \text{if } \Psi_N(i) = 1 \end{cases}_{n \times 1}$$

$$\Psi_N = [\Psi_N(1), \cdots, \Psi_N(n)]^T = \frac{1}{N}\sum_{l=1}^{N} S(l)$$

## 3.2 Asymptotically Efficient Non-Truncated Identification Algorithm

Denote $\Phi\hat{\theta}_N = [\gamma_1, \gamma_2, \cdots, \gamma_n]^T$. With reference to the estimator (3.2) for single-parameter system (3.3), we can easily find that $\gamma_i$ is the estimate of $\phi_i^T \theta$ given by estimator (3.3). By Theorem 3.2 and 3.3, we have

$$E(\gamma_i - \phi_i^T \theta) = O\left(\frac{1}{N}\right)$$

and

$$E(\gamma_i - \phi_i^T \theta)^2 = \frac{F(C - \phi_i^T \theta)(1 - F(C - \phi_i^T \theta))}{Nf(C - \phi_i^T \theta)^2} + O\left(\frac{1}{N^2}\right)$$

Also

$$E(\Phi\hat{\theta}_N - \Phi\theta)(\Phi\hat{\theta}_N - \Phi\theta)^T$$

$$= \begin{bmatrix} E(\gamma_1 - \phi_1\theta)^2 & E(\gamma_1 - \phi_1\theta)E(\gamma_2 - \phi_2\theta) & \cdots & E(\gamma_1 - \phi_1\theta)E(\gamma_n - \phi_n\theta) \\ E(\gamma_2 - \phi_2\theta)E(\gamma_1 - \phi_1\theta) & E(\gamma_2 - \phi_2\theta)^2 & \cdots & E(\gamma_2 - \phi_2\theta)E(\gamma_n - \phi_n\theta) \\ \vdots & \vdots & & \vdots \\ E(\gamma_n - \phi_n\theta)E(\gamma_1 - \phi_1\theta) & E(\gamma_n - \phi_n\theta)E(\gamma_2 - \phi_2\theta) & \cdots & E(\gamma_n - \phi_n\theta)^2 \end{bmatrix}$$

We obtain

$$\Phi E(\hat{\theta}_N - \theta)(\hat{\theta}_N - \theta)^T \Phi^T = \frac{1}{N}\Lambda + O\left(\frac{1}{N^2}\right)\mathbb{I}_{n\times n}$$

where $\Lambda = \text{diag}\left\{\dfrac{F(C - \phi_1^T\theta)(1 - F(C - \phi_1^T\theta))}{f(C - \phi_1^T\theta)^2}, \cdots, \dfrac{F(C - \phi_n^T\theta)(1 - F(C - \phi_n^T\theta))}{f(C - \phi_n^T\theta)^2}\right\}$,

$\mathbb{I}_{n\times n}$ is $n \times n$ dimensional matrix with all the elements being one. Thus, the mean square error of the estimate is

$$E(\hat{\theta}_N - \theta)(\hat{\theta}_N - \theta)^T = \frac{1}{N}\Phi^{-1}\Lambda(\Phi^T)^{-1} + O\left(\frac{1}{N^2}\right)\Phi^{-1}\mathbb{I}_{n\times n}(\Phi^T)^{-1}$$

By Lemma 3.3, we can get the theorem.

### 3.2.3 Numerical Simulation

**Example 3.1** (single parameter system) A simple example is given to illustrate the identification of a single parameter system. Consider system $y(k) = \theta + d(k)$, where the unknown parameter is 5 and the noise is independent and identically distributed Gaussian noise with mean 0 and standard deviation 5. The output is measured by a set-valued sensor, i.e. $s(k) = \mathbb{I}_{\{y(k) \leqslant C\}}$, where the threshold value $C$ is 4. By using the estimation algorithm given in Eq. (3.3), we can get the estimates shown in Fig. 3.1. From the figure, we can see that the estimate can converge to the true value of the parameter.

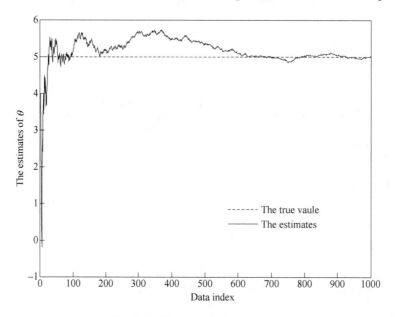

Fig. 3.1  Estimate of the parameter

In this case, the CR lower bound can be calculated. $F(C - \theta) = F(4 - 5) = 0.4207$, $f(C - \theta) = 0.0782$, then the CR lower bounds for $\sigma_{CR}^2(\theta, N) = \dfrac{F(C - \theta)(1 - F(C - \theta))}{Nf(C - \theta)^2} = 39.85/N$. Fig. 3.2 illustrates the simulation results of the CR lower bound and the mean square error of estimation, which shows the mean square error can converge to the CR lower bound.

**Example 3.2** (multi-parameter system)  Consider system $y(k) = \phi^T(k)\theta + d(k)$, where $\theta = [1, 5, 8]^T$, $\phi(k) = [u(k-1), u(k-2), u(k-3)]$. Let $u(k)$ be 3-periodic with $u(1) = 3$, $u(2) = 2$, $u(3) = 3$, then $y(3l + 1) = [1, 2, 3]\theta + d(3l + 1)$, $y(3l + 2) = [3, 1, 2]\theta + d(3l + 2)$, $y(3l + 3) = [2, 3, 1]\theta + d(3l + 3)$, $l = 1, \cdots, N$. Let $Y(l) = [y(3l + 1), y(3l + 2), y(3l + 3)]^T$, we have a $Y(l) = \Phi\theta + D(l)$, where $\Phi = \begin{bmatrix} 1 & 2 & 3 \\ 3 & 1 & 2 \\ 2 & 3 & 1 \end{bmatrix}$, $D(l) = [d(3l + 1), d(3l + 2), d(3l + 3)]^T$.

Sequently, we can get the binary-valued output $S(l) = \mathbb{I}_{\{Y(l) \leqslant C\}}$. According to the esti-

## 3.2 Asymptotically Efficient Non-Truncated Identification Algorithm

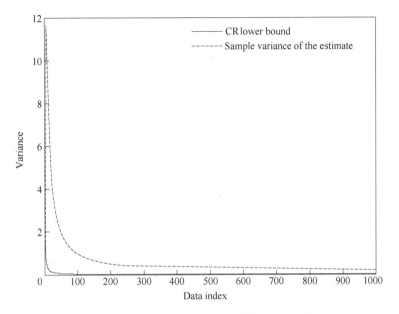

Fig. 3.2 Sample variance vs. CR lower bound

mation algorithm given by Eq. (3.25), we can get the estimate of the parameter. The simulation results are given in Fig. 3.3, which shows that the estimates can converge to the true value of the parameter.

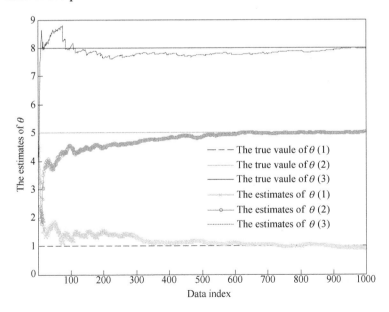

Fig. 3.3 Identification results of the multi-parameter system

## 3.3 Recursive Projection Identification Algorithm

The recursive projection identification algorithm was proposed in literature [38], which proved that the algorithm could converge to the true value of parameters, and the convergence rate was $O\left(\frac{\ln l}{l}\right)$, and the convergence rate was dependent on the prior information of parameters. In this section, we will prove that the algorithm has a faster convergence speed $O(1/l)$, and this speed only depends on the true value of parameters, instead of the prior information of parameters, which is more intuitive than the conditions in [38].

### 3.3.1 Algorithm Design

Considering the system (3.1)

$$\begin{cases} y(k) = \phi^T(k)\theta + d(k) \\ s(k) = I_{y(k) \leq C} \end{cases} \quad k = 1, 2, \cdots \quad (3.26)$$

In order to estimate the parameter in the system (3.1), Ref. [38] proposed the following recursive projection identification algorithm:

$$\begin{cases} \hat{\theta}_l = \Pi_\Theta \left\{ \hat{\theta}_{l-1} + \frac{\beta}{r_l} \phi(l)(F(C - \phi(l)^T \hat{\theta}_{l-1}) - s(l)) \right\} \\ r_l = 1 + \sum_{i=1}^{l} \phi(i)^T \phi(i) \end{cases} \quad (3.27)$$

where $\hat{\theta}_l$ the estimate of parameter $\theta$, $\Pi_\Theta(\cdot)$ is the projection operator mapping from $R^n$ to $\Theta \subset R^n$, which is defined as follows

$$\Pi_\Theta(\zeta) = \operatorname*{argmin}_{\kappa \in \Theta} \| \zeta - \kappa \|, \quad \forall \zeta \in R^n \quad (3.28)$$

In addition, we have two assumptions about unknown parameters and system inputs.

**Assumption 3.3** The unknown parameter $\theta$ belongs to set $\Theta$ which is a bounded convex set, and

$$B = \sup_{v \in \Theta} \| v \|$$

**Assumption 3.4** The inputs $\{\phi(k), k = 1, 2, \cdots\} \in R^n$ satisfy

$$\sup_{k \geq 1} \| \phi(k) \| = M < \infty$$

Besides, there exist a positive integer $T$ and a positive constant $\delta > 0$ such that

$$\sum_{l=k}^{k+T-1} \phi(l)\phi(l)^T \geq \delta I$$

where $I$ is $n\times n$ identity matrix.

**Remark 3.1** ([38], proposition 6)　　The projection operator given by Eq. (3.28) satisfies

$$\|\Pi_\Theta(x_1) - \Pi_\Theta(x_2)\| \leq \|x_1 - x_2\|, \quad \forall x_1, x_2 \in R^n$$

In order to better explain this algorithm, we take $r_l = 1/l$ in algorithm (3.27), and then we get

$$\hat{\theta}_l = \Pi_\Theta\left(\hat{\theta}_{l-1} + \frac{\beta}{l}\phi(l)(F(C - \phi(l)^T\hat{\theta}_{l-1}) - s(l))\right) \quad (3.29)$$

### 3.3.2　Properties of the Algorithm

Denote $\tilde{\theta}_l = \hat{\theta}_l - \theta$ as the estimation error. Then, we can give the results on the estimate. We give a lemma firstly.

**Lemma 3.4** ([38], Lemma 8)　　Under Assumptions 3.1, 3.3 and 3.4, we have

$$\|\tilde{\theta}_i - \tilde{\theta}_j\| \leq \frac{(i-j)\beta B}{j+1}, \quad \text{for } i > j > 0$$

Then, the main results are as follows.

**Theorem 3.6**　　Under Assumptions 3.1, 3.3 and 3.4, the mean square error of the estimate is given as follows

$$E(\tilde{\theta}_l^T \tilde{\theta}_l) = \begin{cases} O\left(\dfrac{1}{l^{2\beta\delta f_B/T}}\right) & \text{if } \beta < \dfrac{T}{2\delta f_B} \\ O\left(\dfrac{\log l}{l}\right) & \text{if } \beta = \dfrac{T}{2\delta f_B} \\ O\left(\dfrac{1}{l}\right) & \text{if } \beta > \dfrac{T}{2\delta f_B} \end{cases}$$

where $f_B = f(|C| + BM)$, $\beta$ is a coefficient in estimation (3.29), and $\delta$ is defined in Assumption 3.4.

**Proof**　　By estimation (3.29), and Remark 3.1, we have

$$\|\tilde{\theta}_l\|^2 \leq \|\tilde{\theta}_{l-1}\|^2 + \frac{2\beta}{l}(F(C - \phi_l^T\hat{\theta}_{l-1}) - s(l))^T \phi_l^T \tilde{\theta}_{l-1} + \frac{\beta^2}{l^2}|\phi_l(F(C - \phi_l^T\hat{\theta}_{l-1}) - s(l))|^2 \quad (3.30)$$

Since $0 < F(\cdot) < 1$ and $s(\cdot)$ equals 0 or 1, we can obtain by Assumption 3.4,

$$E\|\tilde{\theta}_l\|^2 \leq E\|\tilde{\theta}_{l-1}\|^2 + \frac{2\beta}{l}E[(F(C-\phi(l)^T\hat{\theta}_{l-1})-s(l))^T\phi(l)^T\tilde{\theta}_{l-1}] + \frac{\beta^2 M^2}{l^2}$$

(3.31)

Since $\phi_l$ and $\hat{\theta}_{l-1}$ are $\mathcal{F}_{l-1}$ measurable with $\mathcal{F}_{l-1}$ bein $\sigma$-algebra generated by $d(1), \cdots, d(l-1)$, it follows that

$$E[(F(C-\phi(l)^T\hat{\theta}_{l-1})-s(l))^T\phi(l)^T\tilde{\theta}_{l-1}]$$
$$= E\{E[(F(C-\phi(l)^T\hat{\theta}_{l-1})-s(l))^T\phi(l)^T\tilde{\theta}_{l-1} \mid \mathbb{F}_{l-1}]\} \quad (3.32)$$
$$= E[(F(C-\phi(l)^T\hat{\theta}_{l-1})-F(C-\phi(l)^T\theta))^T\phi(l)^T\tilde{\theta}_{l-1}]$$

By the mean value theorem, there exists $\xi_l \in (C-\phi_l^T\hat{\theta}_{l-1}, C-\phi_l^T\theta)$ or $(C-\phi_l^T\theta, C-\phi_l^T\hat{\theta}_{l-1})$ such that

$$F(C-\phi(l)^T\hat{\theta}_{l-1}) - F(C-\phi(l)^T\theta) = -f(\xi_l)\phi(l)^T\tilde{\theta}_{l-1} \quad (3.33)$$

By Assumptions 3.1, 3.2 and 3.4, we have

$$f(\xi_l) \geq f(|C|+BM) \triangleq f_B$$

and consequently, we obtain by Eq. (3.31)

$$E\|\tilde{\theta}_l\|^2 \leq E\|\tilde{\theta}_{l-T}\|^2 - 2\beta f_B \sum_{i=l-T+1}^{l} \frac{1}{i}E(\tilde{\theta}_{i-1}^T\phi(i)\phi(i)^T\tilde{\theta}_{i-1}) + \beta^2 M^2 \sum_{i=l-T+1}^{l} \frac{1}{i^2}$$

(3.34)

Now let's deal with the medial item in the right side of inequality (3.23). Since

$$\tilde{\theta}_{i-1}^T\phi(i)\phi(i)^T\tilde{\theta}_{i-1} = (\tilde{\theta}_{i-1}-\tilde{\theta}_{l-T}+\tilde{\theta}_{l-T})^T\phi(i)\phi(i)^T(\tilde{\theta}_{i-1}-\tilde{\theta}_{l-T}+\tilde{\theta}_{l-T})$$
$$= \tilde{\theta}_{l-T}^T\phi(i)\phi(i)^T\tilde{\theta}_{l-T} + 2(\tilde{\theta}_{i-1}-\tilde{\theta}_{l-T})^T\phi(i)\phi(i)^T\tilde{\theta}_{l-T} + (\phi(i)^T(\tilde{\theta}_{i-1}-\tilde{\theta}_{l-T}))^2$$

It follows that as $l \geq T$

$$\sum_{i=l-T+1}^{l} \frac{1}{i}\tilde{\theta}_{i-1}^T\phi(i)\phi(i)^T\tilde{\theta}_{i-1} \leq -\frac{\delta}{l}\|\tilde{\theta}_{l-T}\|^2 + \frac{4M^2B}{l-T+1}\sum_{i=l-T+1}^{l}\|\tilde{\theta}_{i-1}-\tilde{\theta}_{l-T}\|$$

$$\leq -\frac{\delta}{l}\|\tilde{\theta}_{l-T}\|^2 + O\left(\frac{1}{(l-T+1)^2}\right) \quad (3.35)$$

Thus, we can get

$$-2\beta f_B \sum_{i=l-T+1}^{l} \frac{1}{i}E(\tilde{\theta}_{i-1}^T\phi(i)\phi(i)^T\tilde{\theta}_{i-1}) - \frac{2\delta\beta f_B}{l}E|\tilde{\theta}_{l-T}|^2 + O\left(\frac{1}{(l-T+1)^2}\right)$$

(3.36)

## 3.3 Recursive Projection Identification Algorithm

Taking inequality (3.36) into inequality (3.34) gives

$$E \| \tilde{\theta}_l \|^2 \leq \left(1 - \frac{2\delta\beta f_B}{l}\right) E \| \tilde{\theta}_{l-T} \|^2 + O\left(\frac{1}{(l-T+1)^2}\right) \quad (3.37)$$

Define $\lfloor x \rfloor = \max\{a \in Z | a \leq x\}$, $x \in R$, and $\lceil x \rceil = \min\{a \in Z | a \geq x\}$, $x \in R$. According to inequality (3.37), expanding $\| \tilde{\theta}_l \|^2$ to $\| \tilde{\theta}_{l-\lfloor \frac{l}{T} \rfloor T} \|^2$ yields

$$E \| \tilde{\theta}_l \|^2 \leq \prod_{j=0}^{\lfloor \frac{l}{T} \rfloor - 1} \left(1 - \frac{2\delta\beta f_B}{l - jT}\right) E \| \tilde{\theta}_{l - \lfloor \frac{l}{T} \rfloor T} \|^2 + \sum_{j=1}^{\lfloor \frac{l}{T} \rfloor} \prod_{i=0}^{j-1} \left(1 - \frac{2\delta\beta f_B}{l - iT}\right) O\left(\frac{1}{(l - jT + 1)^2}\right) \quad (3.38)$$

On one hand, the first item in the right side of inequality (3.38) is

$$\sum_{j=0}^{\lfloor \frac{l}{T} \rfloor - 1} \left(1 - \frac{2\delta\beta f_B}{l - jT}\right) \leq \prod_{m = \lceil \frac{l}{T} \rceil - \lfloor \frac{l}{T} \rfloor + 1}^{\lceil \frac{l}{T} \rceil} \left(1 - \frac{2\delta\beta f_B}{Tm}\right)$$

On the other hand, the second item in the right side of inequality (3.38) is

$$\sum_{j=1}^{\lfloor \frac{l}{T} \rfloor} \prod_{i=0}^{j-1} \left(1 - \frac{2\delta\beta f_B}{l - iT}\right) O\left(\frac{1}{(l - jT + 1)^2}\right)$$

$$\leq \sum_{j=1}^{\lfloor \frac{l}{T} \rfloor} \prod_{i=0}^{j-1} \left(1 - \frac{2\delta\beta f_B}{l - iT}\right) O\left(\frac{1}{\left(\lfloor \frac{l}{T} \rfloor - j + \frac{1}{T}\right)^2 T^2}\right)$$

$$\leq O\left(\prod_{i=0}^{\lfloor \frac{l}{T} \rfloor - 1} \left(1 - \frac{2\delta\beta f_B}{l - iT}\right) + \sum_{m=1}^{\lfloor \frac{l}{T} \rfloor - 1} \prod_{p = \lceil \frac{l}{T} \rceil - \lfloor \frac{l}{T} \rfloor + m + 1}^{\lceil \frac{l}{T} \rceil} \left(1 - \frac{2\delta\beta f_B}{pT}\right) \frac{1}{m^2 T^2}\right)$$

By Theorem 2.10, we finally obtain

$$E(\tilde{\theta}_l^T \tilde{\theta}_l) = \begin{cases} O\left(\frac{1}{l^{2\delta\beta f_B/T}}\right) & \text{if } \beta < \frac{T}{2\delta f_B} \\ O\left(\frac{\log l}{l}\right) & \text{if } \beta = \frac{T}{2\delta f_B} \\ O\left(\frac{1}{l}\right) & \text{if } \beta > \frac{T}{2\delta f_B} \end{cases} \quad (3.39)$$

which leads to the theorem.

**Corollary 3.3** For the estimation Eq. (3.29), mean square error of the estimation can achieve convergence rate of $O(1/l)$ if $2\beta\delta f_B > T$.

From the corollary, we can see the convergence rate is faster than that in [[38], Theorem 13]. The condition is weaker than that in [[38], Theorem 13] if the bound of the input matrix satisfies $M > 1$. However, both conditions depend on the bounds of inputs and parameter. In our intuition, the convergence rate of the algorithm should depend on the inputs and the true parameter instead of the boundary values. Thus, we give the following theorem, which is exactly in line with our intuition.

**Theorem 3.7** Under Assumptions 3.1, 3.3 and 3.4, the mean square error of estimation (3.29) can be given by

$$E(\tilde{\theta}_l^T \tilde{\theta}_l) = O(1/l)$$

If $f(C - \phi^T(k)\theta) \geq \underline{f} > \dfrac{T}{2\beta\delta}$ holds for $k = 1, 2, \cdots$, where $\beta$ and $\delta$ are defined in the same as that in Theorem 3.6.

**Proof** Here, we calculate the mean square error $E \| \tilde{\theta}_l \|^2$ by using the higher moment $E \| \tilde{\theta}_l \|^4$. By Eq. (3.31), Eq. (3.32), and Eq. (3.33), we have

$$E \| \tilde{\theta}_l \|^2 \leq E\left[\tilde{\theta}_{l-1}^T\left(I - \frac{2\beta f(\xi_l)}{l}\phi(l)\phi_l^T\right)\tilde{\theta}_{l-1}\right] + O\left(\frac{1}{l^2}\right) \quad (3.40)$$

Sequentially,

$$E \| \tilde{\theta}_l \|^2 \leq E\left[\tilde{\theta}_{l-1}^T\left(I - \frac{2\beta f(C - \phi(l)^T\theta)}{l}\phi(l)\phi(l)^T\right)\tilde{\theta}_{l-1}\right] +$$

$$\frac{2\beta}{l}E[\tilde{\theta}_{l-1}^T\phi(l)(f(C - \phi(l)^T\theta) - f(\xi_l))\phi(l)^T\tilde{\theta}_{l-1}] + O\left(\frac{1}{l^2}\right) \quad (3.41)$$

Due to the mean value theorem, we have

$$f(C - \phi_l^T\theta) - f(\xi_l) = f'(\nu)(C - \phi_l^T\theta - \xi_l)$$

where $\nu \in (C - \phi_l^T\theta, \xi_l)$ or $(\xi_l, C - \phi_l^T\theta)$. Since $\phi(l)$, $\theta$ and $\hat{\theta}$ are bounded, there exists a constant $D$ such that

$$\| f(C - \phi(l)^T\theta) - f(\xi_l) \| \leq \| f'(\nu)\phi(l)^T\tilde{\theta}_{l-1} \| \leq D \| \tilde{\theta}_{l-1} \|$$

Thus

$$E[\tilde{\theta}_{l-1}^T\phi(l)(f(C - \phi_l^T\theta) - f(\xi_l))\phi(l)^T\tilde{\theta}_{l-1}]$$

$$\leq DM^2 E \| \tilde{\theta}_{l-1} \|^3 \leq DM^2 \sqrt{E \| \tilde{\theta}_{l-1} \|^2 E \| \tilde{\theta}_{l-1} \|^4}$$

Then, by Eq. (3.35) we can get

$$E \| \tilde{\theta}_l \|^2 \leq E \| \tilde{\theta}_{l-T} \|^2 - 2\beta \underline{f} \sum_{i=l-T+1}^{l} \frac{1}{i} E(\tilde{\theta}_{i-1}^T \phi(i) \phi_i^T \tilde{\theta}_{i-1}) +$$

## 3.3 Recursive Projection Identification Algorithm

$$2\beta DM^2 \sum_{i=l-T+1}^{l} \frac{1}{i}\sqrt{\|\tilde{\theta}_{i-1}\|^2 E \|\tilde{\theta}_{i-1}\|^4}$$

$$\leq \left(1 - \frac{2\delta\beta f}{l}\right) E\|\tilde{\theta}_{l-T}\|^2 + O\left(\frac{1}{l-T+1}\sqrt{\|\tilde{\theta}_{l-T}\|^2 E\|\tilde{\theta}_{l-T}\|^4}\right) \tag{3.42}$$

Denoting $= \dfrac{\beta\delta f_B}{T}$, we have by Theorem 3.6

$$E\|\tilde{\theta}_l\|^2 = \begin{cases} O\left(\dfrac{1}{l}\right) & \text{if } 2\mu > 1 \\[6pt] O\left(\dfrac{\log l}{l}\right) & \text{if } 2\mu = 1 \\[6pt] O\left(\dfrac{1}{l^{2\mu}}\right) & \text{if } 2\mu < 1 \end{cases} \tag{3.43}$$

Since $\mu$ is a given constant, there exists a positive integer $p$ such that

$$2(p-1)\mu \leq 1, \quad 2p\mu > 1$$

In the following, we will use the property of $E\|\tilde{\theta}_l\|^2$, $E\|\tilde{\theta}_l\|^4$, $\cdots$, $E\|\tilde{\theta}_l\|^{2p}$ to prove the theorem. The main idea of the proof can be given as Fig. 3.4. To explain the detailed proof, we take $p = 1, 2, 3$ for examples, and then give the proof for any $p \geq 4$

(1) If $p = 1$, then $2\mu > 1$. By (3.43), we have

$$E(\tilde{\theta}_l^T \tilde{\theta}_l) = O(1/l)$$

(2) If $p = 2$, then

$$2\mu \leq 1, \quad 4\mu > 1$$

|  | step 1 | step 2 | step 3 | $\cdots$ | step $p-1$ | step $p$ |
|---|---|---|---|---|---|---|
| $E\tilde{\theta}_l^2$ | $O\left(\dfrac{1}{l^{2\mu}}\right)$ | $\to O\left(\dfrac{1}{l^{\left(4-\frac{1}{2^{r-1}}\right)\mu}}\right)$ | $\to O\left(\dfrac{1}{l^{\left(6-\frac{2}{2^{r-1}}\right)\mu}}\right)$ | $\cdots$ | $O\left(\dfrac{1}{l^{\left(2(p-1)-\frac{p-2}{2^{r-1}}\right)\mu}}\right)$ | $\to O\left(\dfrac{1}{l}\right)$ |
| $E\tilde{\theta}_l^4$ | $O\left(\dfrac{1}{l^{4\mu}}\right)$ | $\to O\left(\dfrac{1}{l^{\left(6-\frac{1}{2^{r-1}}\right)\mu}}\right)$ | $\to O\left(\dfrac{1}{l^{\left(8-\frac{2}{2^{r-1}}\right)\mu}}\right)$ | $\cdots$ | $O\left(\dfrac{1}{l^{\left(2p-\frac{p-2}{2^{r-1}}\right)\mu}}\right)$ |  |
| $E\tilde{\theta}_l^6$ | $O\left(\dfrac{1}{l^{6\mu}}\right)$ | $\to O\left(\dfrac{1}{l^{\left(8-\frac{1}{2^{r-1}}\right)\mu}}\right)$ | $\to O\left(\dfrac{1}{l^{\left(10-\frac{2}{2^{r-1}}\right)\mu}}\right)$ | $\cdots$ |  |  |
| $\vdots$ | $\vdots$ | $\vdots$ | $\vdots$ | $\ddots$ |  |  |
| $E\tilde{\theta}_l^{2(p-1)}$ | $O\left(\dfrac{1}{l^{2(p-1)\mu}}\right)$ | $\to O\left(\dfrac{1}{l^{\left(2p-\frac{1}{2^{r-1}}\right)\mu}}\right)$ |  |  |  |  |
| $E\tilde{\theta}_l^{2p}$ | $O\left(\dfrac{1}{l^{2p\mu}}\right)$ |  |  |  |  |  |

Fig. 3.4 Main idea for proving Theorem 3.7

Under this condition, we use the properties of $E \|\tilde{\theta}_{l-1}\|^2$ and $E \|\tilde{\theta}_{l-1}\|^4$ to prove the theorem. Next, the cases $2\mu = 1$ and $2\mu < 1$ will be discussed respectively.

If $2\mu = 1$, we have by (3.43) that there exists a constant $0 < \rho < 1/2$ such that

$$E \|\tilde{\theta}_{l-1}\|^2 = O\left(\frac{1}{l^{1-\rho}}\right) \tag{3.44}$$

By estimation Eq. (3.29) and Remark 3.1, we have

$$\|\tilde{\theta}_l\|^4 \leq \left\|\tilde{\theta}_{l-1} + \frac{\beta}{l}\phi(l)(F(C - \phi(l)^T\hat{\theta}_{l-1}) - s(l))\right\|^4$$

Similar to the proof of Theorem 3.6, we can obtain

$$E \|\tilde{\theta}_l\|^4 \leq \left(1 - \frac{4\delta\beta f_B}{l}\right) E \|\tilde{\theta}_{l-T}\|^4 + O\left(\frac{1}{l^{3-\rho}}\right) \tag{3.45}$$

Similar to the management of equations (3.37) ~ (3.39), we can be obtain by Theorem 2.10 since $4\mu = 2$

$$E \|\tilde{\theta}_l\|^4 \leq O\left(\frac{1}{l^2}\right) + O\left(\frac{1}{l^2} \sum_{m=1}^{\lfloor\frac{l}{T}\rfloor - l} \frac{1}{m^{1-\rho}}\right) = O\left(\frac{1}{l^{2-\rho}}\right) \tag{3.46}$$

Taking Eq. (3.44) into Eq. (3.46), we can obtain

$$E \|\tilde{\theta}_l\|^2 \leq \left(1 - \frac{2\beta\delta f}{l}\right) E \|\tilde{\theta}_{l-T}\|^2 + O\left(\frac{1}{(l - T + 1)^2}\right)$$

$$= \left(1 - \frac{2\beta\delta f}{l}\right) E \|\tilde{\theta}_l\|^2 + O\left(\frac{1}{l^2}\right)$$

$$\leq \prod_{j=0}^{\lfloor\frac{l}{T}\rfloor - 1} \left(1 - \frac{2\beta\delta f}{l - jT}\right) E \|\tilde{\theta}_{l-\lfloor\frac{l}{T}\rfloor T}\|^2 +$$

$$\sum_{j=1}^{\lfloor\frac{l}{T}\rfloor} \prod_{i=0}^{j-1} \left(1 - \frac{2\beta\delta f}{l - iT}\right) O\left(\frac{1}{(l - jT + 1)^2}\right)$$

Since $2\beta\delta f > T$, we have by Theorem 2.10

$$E \|\tilde{\theta}_l\|^2 = O\left(\frac{1}{l}\right)$$

If $2\mu < 1$, we have by Eq. (3.43)

$$E \|\tilde{\theta}_{l-1}\|^2 = O\left(\frac{1}{l^{2\mu}}\right) \tag{3.47}$$

Similar to Eq. (3.45) ~ Eq. (3.46), we can obtain

### 3.3 Recursive Projection Identification Algorithm

$$E|\tilde{\theta}_l|^4 \leq \prod_{j=0}^{\lfloor \frac{l}{T} \rfloor - 1}\left(1 - \frac{4\beta\delta f_B}{l - jT}\right) E\|\tilde{\theta}_{l-\lfloor \frac{l}{T} \rfloor T}\|^4 +$$

$$\sum_{j=0}^{\lfloor \frac{l}{T} \rfloor} \prod_{i=0}^{j-1}\left(1 - \frac{4\beta\delta f_B}{l - iT}\right) O\left(\frac{1}{(l - jT + 1)^{2+2\mu}}\right) \quad (3.48)$$

$$= O\left(\frac{1}{l^{4\mu}}\right)$$

Taking Eq. (3.47) and Eq. (3.48) into Eq. (3.42), we can get

$$E\|\tilde{\theta}_l\|^2 \leq \prod_{j=0}^{\lfloor \frac{l}{T} \rfloor - 1}\left(1 - \frac{2\beta\delta f}{l - jT}\right) E\|\tilde{\theta}_{l-\lfloor \frac{l}{T} \rfloor T}\|^2 +$$

$$\sum_{j=1}^{\lfloor \frac{l}{T} \rfloor} \prod_{i=0}^{j-1}\left(1 - \frac{2\beta\delta f}{l - iT}\right) O\left(\frac{1}{(l - jT + 1)^{3\mu + 1}}\right) + \quad (3.49)$$

$$\sum_{j=1}^{\lfloor \frac{l}{T} \rfloor} \prod_{i=0}^{j-1}\left(1 - \frac{2\beta\delta f}{l - iT}\right) O\left(\frac{1}{(l - jT + 1)^2}\right)$$

Similar to the management of (3.37) ~ (3.39), we can obtain the follows:

(1) If $3\mu \geq 1$, then $E\|\tilde{\theta}_l\|^2 = O\left(\frac{1}{l}\right)$;

(2) If $3\mu < 1$, then

$$E\|\tilde{\theta}_l\|^2 = O\left(\frac{1}{l^{3\mu}}\right) \quad (3.50)$$

Taking Eq. (3.48) and Eq. (3.50) into Eq. (3.42), we have

$$E\|\tilde{\theta}_l\|^2 \leq \left(1 - \frac{2\beta\delta f}{l}\right) E\|\tilde{\theta}_{l-T}\|^2 + O\left(\frac{1}{l^{1+2\mu+3\mu/2}}\right) = \begin{cases} O\left(\frac{1}{l}\right) & \text{if } \frac{7\mu}{2} \geq 1 \\ O\left(\frac{1}{l^{7\mu/2}}\right) & \text{if } \frac{7\mu}{2} < 1 \end{cases}$$

(3.51)

Taking Eq. (3.48) and Eq. (3.51) into Eq. (3.42) again, we can get a new convergence rate $\|\tilde{\theta}_l\|^2$. Repeating this process for $r$ times, we can get

$$E\|\tilde{\theta}_l\|^2 \leq \begin{cases} O\left(\frac{1}{l}\right) & \text{if } \left(4 - \frac{1}{2^{r-1}}\right)\mu \geq 1 \\ O\left(\frac{1}{l^{\left(4 - \frac{1}{2^{r-1}}\right)\mu}}\right) & \text{if } \left(4 - \frac{1}{2^{r-1}}\right)\mu < 1 \end{cases} \quad (3.52)$$

Since $4\mu > 1$, there exists a positive integer $r^*$ such that $\left(4 - \dfrac{1}{2^{r^*-1}}\right)\mu > 1$. Repeating the process for $r^*$ times, we can obtain

$$E \| \tilde{\theta}_l \|^2 = O\left(\dfrac{1}{l}\right)$$

(3) If $p = 3$, then

$$4\mu \leqslant 1, \quad 6\mu > 1$$

Under this condition, we use the properties of $E \| \tilde{\theta}_l \|^2$, $E \| \tilde{\theta}_l \|^4$ and $E \| \tilde{\theta}_l \|^6$ to prove the theorem. The proof can be divided into the following three steps.

(**Step 3.1**) Since $4\mu \leqslant 1$, then $2\mu < 1$. By Eq. (3.43) and Eq. (3.48), we have

$$E \| \tilde{\theta}_{l-1} \|^2 = O\left(\dfrac{1}{l^{2\mu}}\right)$$

and

$$E \| \tilde{\theta}_{l-1} \|^4 = O\left(\dfrac{1}{l^{4\mu}}\right) \tag{3.53}$$

By estimation Eq. (3.29) and Remark 3.1, we have

$$\| \tilde{\theta}_l \|^6 \leqslant \left\| \tilde{\theta}_{l-1} + \dfrac{\beta}{l}\phi(l)(F(C - \phi(l)^T \hat{\theta}_{l-1}) - s(l)) \right\|^6$$

Similar to Eq. (3.48), we have

$$E \| \tilde{\theta}_l \|^6 = O\left(\dfrac{1}{l^{6\mu}}\right) \tag{3.54}$$

(**Step 3.2**) Rewriting $E \| \tilde{\theta}_l \|^2$ by using $E \| \tilde{\theta}_{l-1} \|^4$ gives Eq. (3.42). Taking $E \| \tilde{\theta}_{l-1} \|^4 = O\left(\dfrac{1}{l^{4\mu}}\right)$ and $E \| \tilde{\theta}_{l-1} \|^2 = O\left(\dfrac{1}{l^{2\mu}}\right)$ into (3.42), we can get a new convergence rate $E \| \tilde{\theta}_{l-1} \|^2$. Taking the new and $E \| \tilde{\theta}_{l-1} \|^4 = O\left(\dfrac{1}{l^{4\mu}}\right)$ into Eq. (3.42) gives another new. Repeating the process for $r_1$ times, we can get (3.52). Since $4\mu < 1$, we have $\left(4 - \dfrac{1}{2^{r_1-1}}\right)\mu < 1$ and

$$E \| \tilde{\theta}_l \|^2 = O\left(\dfrac{1}{l^{\left(4 - \frac{1}{2^{r_1-1}}\right)\mu}}\right) \tag{3.55}$$

Similar to Eq. (3.40) and Eq. (3.41), by Eq. (3.54) and every new convergence rate $\| \tilde{\theta}_l \|^4$, and repeating this process for $r_1$ times, we can get the following result since $6\mu < 8\mu \leqslant 2$.

## 3.3 Recursive Projection Identification Algorithm

$$E \| \tilde{\theta}_l \|^4 = O\left(\frac{1}{l^{\left(6-\frac{1}{2^{r_1-1}}\right)\mu}}\right) \quad (3.56)$$

(**Step 3.3**) Taking Eq. (3.55) and Eq. (3.56) into the second item of Eq. (3.42), we can get a new convergence rate of $E \| \tilde{\theta}_l \|^2$. Taking the new convergence rate into Eq. (3.42), we can get another new convergence rate. Repeating the process for $r_2$ times, we can get

$$E|\tilde{\theta}_l|^2 = \begin{cases} O\left(\frac{1}{l}\right) & \text{if } \left(6 - \frac{1}{2^{r_1-1}} - \frac{1}{2^{r_2-1}}\right)\mu \geq 1 \\ O\left(\frac{1}{l^{\left(6-\frac{1}{2^{r_1-1}}-\frac{1}{2^{r_2-1}}\right)\mu}}\right) & \text{if } \left(6 - \frac{1}{2^{r_1-1}} - \frac{1}{2^{r_2-1}}\right)\mu < 1 \end{cases}$$

Since $6\mu > 1$, there exists a positive integers $r^*$ such that $\left(6 - \frac{2}{2^{r^*-1}}\right)\mu > 1$. Let $r_1 = r^*$ and $r_2 = r^*$, then

$$E \| \tilde{\theta}_l \|^2 = O\left(\frac{1}{l}\right)$$

(4) For any $p \geq 4$, we have

$$2\mu < 1, \quad 4\mu < 1, \quad \cdots, \quad 2(p-1)\mu \leq 1, \quad 2p\mu > 1$$

The proof can be divided into $p$ steps.

(**Step 4.1**) Similar to (Step 3.1), we can get the following results by $2\mu < 1$

$$E \| \tilde{\theta}_{l-1} \|^2 = O\left(\frac{1}{l^{2\mu}}\right), \quad \cdots, \quad E \| \tilde{\theta}_{l-1} \|^{2(p-1)} = O\left(\frac{1}{l^{2(p-1)\mu}}\right)$$

$$E \| \tilde{\theta}_l \|^{2p} \leq \left(1 - \frac{2p\delta\beta f_B}{l}\right) E \| \tilde{\theta}_{l-T} \|^{2p} + O\left(\frac{1}{(l-T+1)^2} E \| \tilde{\theta}_{l-T} \|^{2p-1}\right) \leq O\left(\frac{1}{l^{2p\mu}}\right)$$

(**Step 4.2**) Rewriting $E \| \tilde{\theta}_l \|^{2q}$ by using $E \| \tilde{\theta}_{l-1} \|^{2(q+1)}$, $q = 1, 2, \cdots, p-1$ gives

$$E \| \tilde{\theta}_l \|^{2q} \leq \left(1 - \frac{2q\beta\delta f}{l}\right) E \| \tilde{\theta}_{l-T} \|^{2q} + O\left(\frac{1}{l-T+1}\sqrt{E \| \tilde{\theta}_{l-T} \|^{2q} E \| \tilde{\theta}_{l-T} \|^{2(q+1)}}\right) \quad (3.57)$$

Applying the results on $E \| \tilde{\theta}_{l-1} \|^{2q}$ and $E \| \tilde{\theta}_{l-1} \|^{2(q+1)}$ in (Step 4.1) to the above inequality we can get a new convergence rate of $E \| \tilde{\theta}_l \|^{2q}$. By repeating the process for $r$ times, we can get the convergence rate as follows.

$$E\|\tilde{\theta}_l\|^{2q} = O\left(\frac{1}{l^{\left(2(q+1) - \frac{1}{2^{r-1}}\right)\mu}}\right), \quad q = 1, 2, \cdots, p - 1 \quad (3.58)$$

(**Step 4.3**) Applying the results (3.58) to inequality (3.57) again, we can get the new convergence rate by repeating the process for $r$ times

$$E\|\tilde{\theta}_l\|^{2q} = O\left(\frac{1}{l^{\left(2(q+2) - \frac{2}{2^{r-1}}\right)\mu}}\right) \quad q = 1, 2, \cdots, p - 2$$

(Setp 4.4), (Step 4.5), $\cdots$ (Step 4.p-1) are similar to (Step 4.3). From step (4.p-1), we can get

$$E\|\tilde{\theta}_l\|^2 = O\left(\frac{1}{l^{\left(2(p-1) - \frac{p-2}{2^{r-1}}\right)\mu}}\right) \quad (3.59)$$

and

$$E\|\tilde{\theta}_l\|^4 = O\left(\frac{1}{l^{\left(2p - \frac{p-2}{2^{r-1}}\right)\mu}}\right) \quad (3.60)$$

Next, we will give the last step.

(**Step 4.p**) Taking Eq. (3.59) and Eq. (3.60) into the second item of inequality (3.42), we can get a new convergence rate of $E\|\tilde{\theta}_l\|^2$. Repeating the process for $r$ times, we can get

$$E\|\tilde{\theta}_l\|^2 = \begin{cases} O\left(\dfrac{1}{l}\right) & \text{if } \left(2p - \dfrac{p-1}{2^{r-1}}\right)\mu \geqslant 1 \\ O\left(\dfrac{1}{l^{\left(2p - \frac{p-1}{2^{r-1}}\right)\mu}}\right) & \text{if } \left(2p - \dfrac{p-1}{2^{r-1}}\right)\mu < 1 \end{cases}$$

Since $2p\mu > 1$, there exists a positive integers $r^*$ such that $\left(2p - \dfrac{p-1}{2^{r^*-1}}\right)\mu > 1$. Let $r = r^*$, then we can obtain

$$E\|\tilde{\theta}_l\|^2 = O\left(\frac{1}{l}\right)$$

The theorem is proved.

### 3.3.3 Numerical Simulation

Consider a first-order system: $y_t = \phi_t \theta + d_t$, whose quantized output is $s_t = I_{\{y_t \leqslant C\}}$, where $C = 0.5$ is a known constant threshold, the system noise $d_t$ is from $N(0, 1)$, the true parameter $\theta = 0.25$ is unknown, but the prior information $|\theta| \leqslant 1$ can be obtained, the inputs follow $\dfrac{1}{4} \leqslant \phi_t \leqslant 1$ and are randomly generated in the interval $[1/4, 1]$. By using the recursive projection algorithm (3.29), we give the simulation results in Fig. 3.5. We can see that the estimate converges to the true parameter $\theta$.

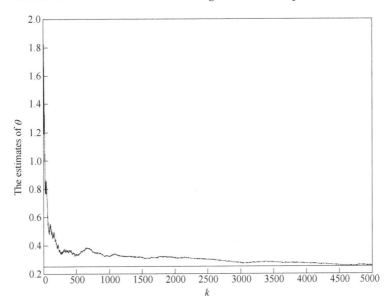

Fig. 3.5 Simulation results of the recursive projection identification algorithm

## 3.4 Notes

In this chapter, two identification algorithms for binary valued systems are introduced: non-truncated identification algorithm and recursive projection identification algorithm. The non-truncated identification algorithm is based on the classical empirical measure method. Compared with the empirical measure method, it does not require the prior information of the parameters, and it has wider application. In this chapter, the identification of single-parameter and multi-parameter systems is considered respectively, and the asymptotic efficiency of the algorithm is proved under periodic input conditions. Recursive projection identification algorithm does not require the input to be periodic, and it is an online identification algorithm. For this algorithm, we prove that the convergence rate is faster than that in the existing literature, and the convergence rate depends on the true value of parameter rather than the prior information given in the existing literature.

Based on this work, there are many interesting problems that need to be further studied. For examples, could these two algorithms be applied to set-valued systems with multiple thresholds? Could the mean square error of recursive projection algorithm converge to the CR lower bound?

# 4 Consensus with Binary-Valued Measurements under Undirected Topology

The recent years have witnessed a lot of researches on cooperative control of multi-agent systems. The common feature of these systems is that the agents are required to cooperate with others to achieve a group objective. Cooperative control for multi-agent systems can be categorized as either centralized control in which a powerful central station is available to control a group of agents, or distributed control which does not require the existence of the central station[39]. Due to inevitable practical constraints, distributed control is more promising and can be applied to wide areas, such as load balancing of multi-processors[40], coordinated flocking of mobile vehicles[41~44], cooperative control of unmanned ground, aerial and underwater vehicles[6,8,45,46], and attitude alignment of clusters of UAVs[47~49].

The problem of consensus is one of the most basic and important distributed coordinated control of the multi-agent system, and it is also the basis of many distributed control and parameter estimation problems. Quantized consensus emerges from the sensor networks with the development of digital communication. But most existing literatures require quantization level bigger than three to ensure consensus. The quantization level is smaller, the information obtained is less and the consensus control will be more difficult.

In this chapter, we study the consensus problem of multi-agent systems with binary-valued and noisy measurements. In the system, each agent can only obtain the binary-valued information from its neighbors. That is, each agent can only know whether the state of its neighbor is greater than a certain threshold. How to design control algorithm by using the binary-valued information to achieve consensus? In the case of accurate state communication, consensus control is designed by the states of its neighbors. In this chapter, the communication is binary-valued, where the neighbors' states are unknown. One natural idea is to replace the true states with the estimates of neighbors' states. Therefore, we use the two identification methods given in Chapter 3 to estimate neighbors' states, and then design two kinds of consensus algorithms based on the estimates.

The first consensus algorithm is the two-time-scale consensus algorithm, which is given by the non-truncated identification algorithm. The other is the recursive projection consensus algorithm, which is given by the recursive projection identification algorithm. In the two-time-scale algorithm, the states are required to keep unchanged for a period of time. During this time, the states are estimated by the non-truncated identification algorithm. The control is designed based on the estimates, which is performed only at some skipping times. In order to avoid waiting a period of time, we give the recursive projection consensus algorithm. The identification algorithm is online, which makes the control perform at every moment. Finally, we prove both of the consensus algorithms can achieve consensus and the consensus speeds are given respectively.

The structure of this chapter is as follows. Section 4.1 is the problem formulation. Section 4.2 introduces the two-time-scale consensus algorithm. The convergence of the algorithm is proved and the convergence rate is given. Section 4.3 introduces the recursive projection consensus algorithm. The convergence and convergence rate of the algorithm are given. Section 4.4 gives some notes.

## 4.1 Problem Formulation

Consider a multi-agent system of $n$ nodes, given by

$$x_i(t+1) = x_i(t) + u_i(t), \quad i = 1, \cdots, n \tag{4.1}$$

for $t = 1, 2, \cdots$, which can be written in a vector form:

$$x(t+1) = x(t) + u(t) \tag{4.2}$$

with $x(t) = [x_1(t), \cdots, x_n(t)]^T$ and $u(t) = [u_1(t), \cdots, u_n(t)]^T$, where $x_i(t)$ is the state of the $i$th node at time $t$, $u_i(t)$ is the control of node $i$ at time $t$.

The agents in the system (4.1) are distributed with a spatial structure which is described by an undirected graph $G = (N, E)$ consisting of a set of nodes $N = \{1, 2, \cdots, n\}$ and a set of edges $E \subset N \times N$. We denote each edge as an unordered pair $(i, j)$ where $i \neq j$. If $(i, j) \in E$, then node $i$ (resp, node $j$) is the neighbor of node $j$ (resp, node $i$), denoted as $i \in N_j$ (resp, $j \in N_i$).

For each node $i$, it can only get the binary-valued measurements from its neighbors $j \in N_i$:

$$\begin{cases} y_{ij}(t) = x_j(t) + d_{ij}(t) \\ s_{ij}(t) = I_{\{y_{ij}(t) \leq C\}} = \begin{cases} 1 & \text{if } y_{ij}(t) \leq C \\ 0 & \text{if } y_{ij}(t) > C \end{cases} \end{cases} \tag{4.3}$$

where $x_j(t)$ is the state of node $j$ at the time $t$, $d_{ij}(t)$ is the communication noise,

$y_{ij}(t)$ is the output data which cannot be measured, and $s_{ij}(t)$ is the observed binary-valued data.

For the spatial structure $G$ and the noises, we have the following assumptions:

**Assumption 4.1** The spatial structure $G$ is connected.

**Assumption 4.2** The noises $\{d_{ij}(t), (i, j) \in G, t=1, 2, \cdots\}$ are independent with respect to $i$, $j$, $t$ and identically normally distributed. The distribution function and the associated density function are $F(\cdot)$ and $f(x) = dF(x)/dx \neq 0$, respectively.

The objective is to design the control $u_i(t)$, $i=1, \cdots, n$, to achieve consensus for the multi-agent system (4.1) with binary-valued communication (4.3).

**Definition 4.1** (Weak Consensus) The agents are said to reach weak consensus if $E|x_i(t)|^2 < \infty$, $t \geq 0$, $i \in N$, and $\lim_{t \to \infty} E|x_i(t) - x_j(t)|^2 = 0$ for all distinct $i, j \in N$.

**Definition 4.2** (Mean Square Consensus) The agents are said to reach mean square consensus if $E|x_i(t)|^2 < \infty$, $t \geq 0$, $i \in N$, and there exists a random variable $x^*$ such that $\lim_{t \to \infty} E|x_i(t) - x^*|^2 = 0$ for all $i \in N$.

## 4.2 Two-Time-Scale Consensus

### 4.2.1 Estimation

To achieve consensus, each agent needs to design the consensus control by using the states of its neighbors. But the states of the neighbors' are unknown, so each agent should estimate the states of its neighbors and then design the consensus control. Since each agent can only get binary-valued information from its neighbors, it should accumulate the information for a period of time to get enough information to estimate its neighbors' states.

As a result, we propose a protocol alternating estimation with control: each agent estimates its neighbors' states, then designs the control, by which the state will be updated. And then the process of estimation alternating with control will be repeated. As time goes, the estimation will be more precise and the system will achieve consensus after controlling many times. The process can be shown as the following Fig. 4.1.

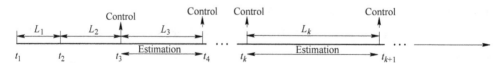

Fig. 4.1 The structure of two-time-scale consensus algorithm

Agent $i(i=1, \cdots, n)$ estimates the states of its neighbors at the time $t_k$ for a holding time $L_k$ during which the states of the agents keep constant. Based on the estimations of

## 4.2 Two-Time-Scale Consensus

its neighbors' states, each agent $i(i=1, \cdots, n)$ designs the control $u_i(t)$ at the time $t = t_{k+1} - 1 = t_k + L_k - 1$ by the averaging rule with $1/k$ step size. Finally, the states of the nodes will be updated by system (4.1) at time $t_{k+1} - 1$ and the process will be repeated: estimation, control, estimation, control.

Since the states keep constant for the holding time $L_k$, an estimation is carried out on the interval $[t_k, t_{k+1} - 1)$. Now let's focus on the binary-valued measurements on the interval $[t_k, t_{k+1} - 1)$.

$$\begin{cases} y_{ij}(l) = x_j(t_k) + d_{ij}(l) \\ s_{ij}(l) = I_{\{y_{ij}(l) \leq C\}} \end{cases} \quad j \in N_i, \ l = t_k, \cdots, t_{k+1} - 1 \quad (4.4)$$

Under Assumption 4.2, each link can be studied individually, independent of others. As a result, for notational simplicity, we will omit the link specification $(i, j)$ in communication system (4.4) and study a general link.

$$\begin{cases} y(l) = \theta_k + d(l) \\ s(l) = I_{\{y(l) \leq C\}} \end{cases} \quad l = t_k, \cdots, t_{k+1} - 1 \quad (4.5)$$

where $\theta_k$ corresponds to the state $x_j(t_k)$ of any node $j$ in (4.4), $\{d(l)\}$ satisfies Assumption 4.2.

The non-truncated identification estimation is

$$\varphi_k = \frac{1}{L_k} \sum_{l=t_k}^{t_{k+1}-1} s(l); \quad \xi_k = \begin{cases} \dfrac{1}{2} & \text{if } \varphi_k = 0 \\ \varphi_k & \text{if } 0 < \varphi_k < 1 \\ \dfrac{1}{2} & \text{if } \varphi_k = 1 \end{cases}$$

$$\hat{\theta}_k = C - F^{-1}(\xi_k) \quad (4.6)$$

For system (3.2), in which the state is a constant, the estimation $\hat{\theta}_N$ has good property as theorem 3.2 says. For system (4.5), the state $\theta_k$ keeps unchanged only for a holding time $L_k$. Is the estimation $\hat{\theta}_k$ calculated by Eq. (4.6) has the same property as Theorem 3.2? The answer is yes if the state changes in a bounded set.

**Theorem 4.1** For system (4.5), the estimation of the state $\theta_k$ is calculated by Eq. (4.6). If the state $\theta_k$ is bounded, the mean square error of the estimation can be given by

$$E(\hat{\theta}_k - \theta_k)^2 = O\left(\frac{1}{L_k}\right) \quad (4.7)$$

where $\hat{\theta}_k$ is the estimation of $\theta_k$, $L_k = t_{k+1} - t_k$ is the holding time.

**Proof** The main idea of the proof is similar to the proof of Theorem 3.2. But some details are different. Now let's see the proof in detail.

By Eq. (4.6), we have

$$E(\hat{\theta}_k - \theta_k)^2 = E(C - \theta_k)^2 - 2E(F^{-1}(\xi_k)(C - \theta_k)) + E(F^{-1}(\xi_k))^2 \quad (4.8)$$

Firstly, let's see the item $E(F^{-1}(\xi_k)(C - \theta_k))$. By the property of conditional expectation, we have

$$E(F^{-1}(\xi_k)(C - \theta_k)) = E[E(F^{-1}(\xi_k)(C - \theta_k) \mid \theta_k)]$$
$$= E[(C - \theta_k)E(F^{-1}(\xi_k)) \mid \theta_k] \quad (4.9)$$

Due to the construction of $\xi_k$, we can obtain

$$E(F^{-1}(\xi_k) \mid \theta_k)$$
$$= F^{-1}\left(\frac{1}{2}\right)P\left(\xi_k = \frac{1}{2} \mid \theta_k\right) + \sum_{l=1}^{L_k-1} F^{-1}\left(\frac{1}{L_k}\right)P\left(\xi_k = \frac{1}{L_k} \mid \theta_k\right)$$
$$= \sum_{l=1}^{L_k-1} F^{-1}\left(\frac{1}{L_k}\right)C_{L_k}^l p_k^l(1 - p_k)^{L_k-l} + F^{-1}\left(\frac{1}{2}\right)(p_k^{L_k} + (1 - p_k)^{L_k})$$

where $p_k = P(s(l) = 1 \mid \theta_k) = F(C - \theta_k)$. Noting that $p_k$ is a random variable, while $p$ in Eq. (3.15) is a constant. Since $\theta_k$ is bounded, there exist positive constants $b$ and $\delta$ such that

$$\delta < b < p_k < 1 - b < 1 - \delta, \text{ a. s.}$$

Dividing $E(F^{-1}(\xi_k) \mid \theta_k)$ into four parts by $\delta$, we have

$$E(F^{-1}(\xi_k) \mid \theta_k) = F^{-1}\left(\frac{1}{2}\right)(p_k^{L_k} + (1 - p_k)^{L_k}) +$$
$$\sum_{\frac{1}{L_k} < \frac{l}{L_k} < \frac{\delta}{2}} F^{-1}\left(\frac{1}{L_k}\right)C_{L_k}^l p_k^l(1 - p_k)^{L_k-l} +$$
$$\sum_{\frac{\delta}{2} \leq \frac{l}{L_k} \leq 1-\frac{\delta}{2}} F^{-1}\left(\frac{1}{L_k}\right)C_{L_k}^l p_k^l(1 - p_k)^{L_k-l} +$$
$$\sum_{1-\frac{\delta}{2} < \frac{l}{L_k} < 1-\frac{1}{L_k}} F^{-1}\left(\frac{1}{L_k}\right)C_{L_k}^l p_k^l(1 - p_k)^{L_k-l} \quad (4.10)$$

Let

$$\tilde{Q}_1 = \sum_{\frac{1}{L_k} < \frac{l}{L_k} < \frac{\delta}{2}} F^{-1}\left(\frac{1}{L_k}\right)C_{L_k}^l p_k^l(1 - p_k)^{L_k-l}$$

## 4.2 Two-Time-Scale Consensus

$$\tilde{Q}_2 = \sum_{1-\frac{\delta}{2} < \frac{l}{L_k} < 1-\frac{l}{L_k}} F^{-1}\left(\frac{1}{L_k}\right) C_{L_k}^l p_k^l (1-p_k)^{L_k-l}$$

$$\tilde{Q}_3 = \sum_{\frac{\delta}{2} \leq \frac{l}{L_k} \leq 1-\frac{\delta}{2}} F^{-1}\left(\frac{1}{L_k}\right) C_{L_k}^l p_k^l (1-p_k)^{L_k-l}$$

Also, due to the boundedness of $p_k$, there exists $d_4 > 0$ such that

$$p_k^{L_k} + (1-p_k)^{L_k} = O(e^{-d_4 L_k})$$

Thus,

$$E(F^{-1}(\xi_k) \mid \theta_k) = \tilde{Q}_1 + \tilde{Q}_2 + \tilde{Q}_3 + O(e^{-d_4 L_k}) \qquad (4.11)$$

We can see $\tilde{Q}_1$, $\tilde{Q}_2$ and $\tilde{Q}_3$ have the same form:

$$F^{-1}(1/L_k) C_{L_k}^l p_k^l (1-p_k)^{L_k-l}$$

$$= F^{-1}(1/L_k) \frac{L_k!}{l!(L_k-l)!} p_k^l (1-p_k)^{L_k-l}$$

$$= \frac{L_k!}{l!(L_k-l)!} \delta^l (1-\delta)^{L_k-l} F^{-1}(1/L_k) \left(\frac{p_k}{\delta}\right)^l \left(\frac{1-p_k}{1-\delta}\right)^{L_k-l}$$

Since

$$\log\left[\left(\frac{p_k}{\delta}\right)^l \left(\frac{1-p_k}{1-\delta}\right)^{L_k-l}\right] = g(p_k) - g(\delta) = g'(v_1)(p_k - \delta) = \frac{L_k(l/L_k - v_1)}{v_1(1-v_1)}(p_k - \delta)$$

where $\delta < v_1 < p_k$, $l/L_k < \frac{\delta}{2}$, $\delta < b < p_k < 1-b < 1-\delta$. Thus, we have

$$\left(\frac{p_k}{\delta}\right)^l \left(\frac{1-p_k}{1-\delta}\right)^{L_k-l} \leq -\frac{\delta(b-\delta)}{2} L_k$$

Denote $\Delta_5 = \frac{\delta(b-\delta)}{2}$, we obtain

$$\left(\frac{p_k}{\delta}\right)^l \left(\frac{1-p_k}{1-\delta}\right)^{L_k-l} \leq e^{-\Delta_5 L_k}$$

Similar to Eq. (3.7), we acquire

$$|\tilde{Q}_1| \leq \max\left\{\frac{\delta L_k}{2} F^{-1}(1/L_k) e^{-\Delta_5 L_k}, \frac{\delta L_k}{2} F^{-1}(\delta/2) e^{-\Delta_5 L_k}\right\}$$

Since $\frac{\delta L_k}{2} F^{-1}(1/L_k) e^{-\Delta_5 L_k} \to 0$, as $k \to \infty$, we have

$$\tilde{Q}_1 = O(e^{-d_5 L_k})$$

where $d_5 = \Delta_5/2$. Analogously, $\tilde{Q}_2 = O(e^{-d_6 L_k})$.

The item $\tilde{Q}_3$ is,

$$\tilde{Q}_3 = \sum_{\frac{\delta}{2} \leq \frac{l}{L_k} \leq 1 - \frac{\delta}{2}} \left( G(p_k) + G'(p_k)(l/L_k - p_k) + \frac{G''(\eta)}{2}(l/L_k - p_k)^2 \right) C_{L_k}^l p_k^l (1 - p_k)^{L_k - l}$$

$$= G(p_k) + O(e^{-d_4 L_k}) + O(e^{-d_5 L_k}) + O(e^{-d_5 L_k}) + O\left(\frac{1}{L_k}\right)$$

$$= C - \theta_k + O\left(\frac{1}{L_k}\right)$$

By Eq. (4.11), we have

$$E(F^{-1}(\xi_k) \mid \theta_k) = C - \theta_k + O\left(\frac{1}{L_k}\right) \tag{4.12}$$

Analogously

$$E((F^{-1}(\xi_k))^2 \mid \theta_k) = (C - \theta_k)^2 + O\left(\frac{1}{L_k}\right) \tag{4.13}$$

Substituting Eq. (4.12) into Eq. (4.9), together with the boundedness of $\theta_k$, yields

$$E(F^{-1}(\xi_k)(C - \theta_k)) = E[(C - \theta_k)((C - \theta_k) + O(1/L_k))]$$
$$= E(C - \theta_k)^2 + O\left(\frac{1}{L_k}\right) \tag{4.14}$$

Secondly, we have by Eq. (4.13)

$$E(F^{-1}(\xi_k))^2 = E[E((F^{-1}(\xi_k))^2 \mid \theta_k)] = E(C - \theta_k)^2 + O\left(\frac{1}{L_k}\right) \tag{4.15}$$

Substituting Eq. (4.14) and Eq. (4.15) into Eq. (4.8), we can get the theorem.

**Remark 4.1** In Theorem 3.2, $\theta$ is a constant parameter. In Theorem 4.1, $\theta_k$ keeps unchanged for the holding time $L_k$, but it may be changed by the control as $k$ changes. If $\theta_k$ is bounded, we can still get the same result as that in Theorem 3.2.

Add the link $(i, j)$ into system (4.5), we can get the system (4.4). For all $j \in N_i$, $i = 1, \cdots, n$, the corresponding estimation $\hat{z}_{ij}(L_k)$ is as follows,

$$\varphi_{ij}(L_k) = \frac{1}{L_k} \sum_{l=t_k}^{t_{k+1}-1} s_{ij}(l)$$

$$\xi_{ij}(L_k) = \begin{cases} \frac{1}{2} & \text{if } \varphi_{ij}(L_k) = 0 \text{ or } 1 \\ \varphi_{ij}(L_k) & \text{if } 0 < \varphi_{ij}(L_k) < 1 \end{cases}$$

## 4.2 Two-Time-Scale Consensus

$$\hat{z}_{ij}(L_k) = C - F^{-1}(\xi_{ij}(L_k)) \tag{4.16}$$

The state $x_j(t_k)$ in Eq. (4.4) corresponds to $\theta_k$ in Eq. (4.5), and the estimation $\hat{z}_{ij}(L_k)$ in Eq. (4.16) corresponds to $\hat{\theta}_k$ in Eq. (4.6). We can get the following results as Theorem 4.1.

**Theorem 4.2** The estimations of the neighbors' states are calculated by Eq. (4.16). If the state of the neighbor $j(j \in N_i, i = 1, \cdots, n)$ is bounded, the mean square error of the estimation will be

$$E(\hat{z}_{ij}(L_k) - x_j(t_k))^2 = O\left(\frac{1}{L_k}\right), \quad \text{as} \quad L_k \to \infty \tag{4.17}$$

where $x_j(t_k)$ is the state of neighbor $j$, $\hat{z}_{ij}(L_k)$ is the estimation of neighbor $j$'s state, $L_k$ is the holding time, $N_i$ is the set of the neighbors of agent $i$.

### 4.2.2 Consensus Control

If the states of its neighbors can be obtained, the states can be updated by the averaging rule to achieve consensus:

$$x_i(t+1) = \frac{1}{|N_i|} \sum_{j \in N_i} x_j(t)$$

where $|N_i|$ is the number of neighbors of agent $i$. The state updating implies that consensus control is

$$u_i(t) = -\frac{1}{|N_i|} \sum_{j \in N_i} (x_i(t) - x_j(t)) \tag{4.18}$$

In this chapter, the states of the neighbors are unknown, and the control is designed only at the skipping time $t_k$, $k = 1, 2, \cdots$. Replacing the states of the neighbors in Eq. (4.18) with the estimate, the control can be given in a stochastic approximation form:

$$\tilde{u}_i(k) = -\frac{1}{k} \sum_{j \in N_i} (x_i(t_k) - \hat{z}_{ij}(L_k)), \quad i = 1, \cdots, n \tag{4.19}$$

By Theorem 4.2, the states need to be bounded. So each agent checks whether the following condition is satisfied,

$$|x_i(t_k) + \tilde{u}_i(k)| < h_i(k+1) \tag{4.20}$$

where $h_i(k+1) = M_i + \sum_{h=1}^{k+1} \frac{1}{h^{1+\gamma}}$, $M_i > |x_i(0)|$, $\gamma > 0$. If the condition Eq. (4.20) is satisfied, $u_i(t_{k+1}-1) = \tilde{u}_i(k)$; If not, $u_i(t_{k+1}-1) = 0$. In a word, the control $u_i(t_{k+1}-1)$ is

$$u_i(t_{k+1} - 1) = -\frac{1}{k} \sum_{j \in N_i} (x_i(t_k) - \hat{z}_{ij}(L_k)) I_{\{|x_i(t_k) + \tilde{u}_i(k)| < h_i(k+1)\}} \tag{4.21}$$

## 4 Consensus with Binary-Valued Measurements under Undirected Topology

By system (4.1), the states will be updated by

$$x_i(t_{k+1}) = x_i(t_k) - \frac{1}{k}\sum_{j \in N_i}(x_i(t_k) - \hat{z}_{ij}(L_k))I_{\{|x_i(t_k)+\bar{u}_i(k)|<h_i(k+1)\}} \quad (4.22)$$

**Remark 4.2** By the updating Eq. (4.22), the state of each node $i=1, \cdots, n$ will satisfy the following inequality,

$$|x_i(t_k)| < h_i(k) = M_i + \sum_{h=1}^{k}\frac{1}{h^{1+\gamma}} \leq M_i + \sum_{h=1}^{\infty}\frac{1}{h^{1+\gamma}}$$

if $x_i(t_0) < M_i$. This implies that the states of the nodes are bounded at any time, and the boundary is $\widetilde{M}_i \triangleq M_i + \sum_{h=1}^{\infty}\frac{1}{h^{1+\gamma}}$ for node $i$.

Denote $z_i(k) = x_i(t_k)$, the state will be updated by Eq. (4.22), which can be written as:

$$z_i(k+1) = z_i(k) - \frac{1}{k}\sum_{j \in N_i}(z_i(k) - \hat{z}_{ij}(L_k))I_{\{|z_i(k)+\bar{u}_i(k)|<h_i(k+1)\}}$$

$$= z_i(k) - \frac{1}{k}\sum_{j \in N_i}(z_i(k) - \hat{z}_{ij}(L_k)) +$$

$$\frac{1}{k}\sum_{j \in N_i}(z_j(k) - \hat{z}_{ij}(L_k))I_{\{|z_i(k)+\bar{u}_i(k)|\geq h_i(k+1)\}}$$

Denote $\hat{z}_{ij}(L_k) - z_j(k) = \varepsilon_{ij}(k)$, and

$$R_{ij}(k) = (z_i(k) - \hat{z}_{ij}(L_k))I_{\{|z_i(k)+\bar{u}_i(k)|\geq h_i(k+1)\}} \quad (4.23)$$

Then $z_i(k+1)$ is actually

$$z_i(k+1) = z_i(k) - \frac{1}{k}\sum_{j \in N_i}(z_i(k) - z_j(k)) + \frac{1}{k}\sum_{j \in N_i}\varepsilon_{ij}(k) + \frac{1}{k}\sum_{j \in N_i}R_{ij}(k)$$

which can be written in a vector form

$$z(k+1) = z(k) - \frac{1}{k}Lz(k) + \frac{1}{k}\varepsilon(k) + \frac{1}{k}R(k) \quad (4.24)$$

where $z(k) = (z_1(k), \cdots, z_n(k))^T$, $L$ is the Laplacian Matrix of the undirected graph $G$, $\varepsilon(k) = (\varepsilon_1(k), \cdots, \varepsilon_n(k))^T$, $\varepsilon_i(k) = \sum_{j \in N_i}\varepsilon_{ij}(k)$, $R(k) = (R_1(k), \cdots, R_n(k))^T$, $R_i(k) = \sum_{j \in N_i}R_{ij}(k)$.

### 4.2.2.1 Convergence

In the general case, the more information each agent gets, the more accurately it estimates. Thus we choose the holding time for estimation to be an increasing function. For simplicity, we take $L_k = k^\alpha$, $\alpha > 0$, then $L_k \to \infty$, as $k \to \infty$.

To give the convergence of the states, we give the following lemmas firstly.

**Lemma 4.1** [[19], Theorem 5] Under Assumption 4.1, we have the assertions:

(1) The null spaces of $L$, $L^2$ and $L^3$ are given by the same one dimensional space, i.e., $N_i = \text{span}\{\mathbb{1}_n\}$, $i = 1, 2, 3$, where one is the $n$ dimensional vector with all elements are 1.

(2) Let

$$0 = \lambda_1 < \lambda_2 \leq \lambda_3 \leq \cdots \leq \lambda_n$$

$$0 = \hat{\lambda}_1 < \hat{\lambda}_2 \leq \hat{\lambda}_3 \leq \cdots \leq \hat{\lambda}_n$$

and

$$0 = \tilde{\lambda}_1 < \tilde{\lambda}_2 \leq \tilde{\lambda}_3 \leq \cdots \leq \tilde{\lambda}_n$$

denote the eigenvalues of $L$, $L^2$ and $L^3$, respectively. There exist positive constants $\beta_1 = \hat{\lambda}_2 \lambda_n^{-1} > 0$ and $\beta_2 = \tilde{\lambda}_n \lambda_2^{-1} > 0$ such that $L^2 \geq \beta_1 L$ and $L^3 \leq \beta_2 L$.

**Lemma 4.2** Let the holding time $L_k = k^\alpha$, we have the following assertion under Assumption 4.2:

$$ER_{ij}^2(k) = O\left(\frac{1}{k^{\alpha-2\gamma}}\right), \quad \text{for all} \quad (i, j) \in E \tag{4.25}$$

where $R_{ij}(k)$ is defined by Eq. (4.23), $\gamma$ is the parameter in the state bound $h_i(k)$ in Eq. (4.20).

**Proof** Since $R_{ij}(k)$ is defined by Eq. (4.23), we have

$$R_{ij}^2(k) = (z_i(k) - z_j(k) - \varepsilon_{ij}(k))^2 I_{\{|z_i(k) + \tilde{u}_i(k)| \geq h_i(k+1)\}}$$

$$\leq 2(z_i(k) - z_j(k))^2 I_{\{|z_i(k) + \tilde{u}_i(k)| \geq h_i(k+1)\}} +$$

$$2\varepsilon_{ij}^2(k) I_{\{|z_i(k) + \tilde{u}_i(k)| \geq h_i(k+1)\}} \tag{4.26}$$

By Remark 4.2 and Theorem 4.2, we have for any $(i, j) \in E$,

$$E(z_i(k) - z_j(k))^2 I_{\{|z_i(k) + \tilde{u}_i(k)| \geq h_i(k+1)\}}$$

$$\leq (\tilde{M}_i + \tilde{M}_j)^2 P\{|z_i(k) + \tilde{u}_i(k)| \geq h_i(k+1)\}$$

and

$$E\varepsilon_{ij}^2(k)I_{\{|z_i(k)+\tilde{u}_i(k)|\geqslant h_i(k+1)\}} \leqslant E\varepsilon_{ij}^2(k) = O\left(\frac{1}{L_k}\right)$$

Let

$$D(k) = P\{|z_i(k) + \tilde{u}_i(k)| \geqslant h_i(k+1)\} \tag{4.27}$$

Taking expectation on both sides of Eq. (4.26) yields

$$ER_{ij}^2(k) \leqslant O(D(k)) + O\left(\frac{1}{L_k}\right) \tag{4.28}$$

Now, let's focus on $D(k)$. Substituting Eq. (4.19) to Eq. (4.27), we have

$$D(k) = P\left\{\left|z_i(k) - \frac{1}{k}\sum_{j\in N_i}(z_i(k) - z_j(k)) + \frac{1}{k}\sum_{j\in N_i}\varepsilon_{ij}(k)\right| \geqslant h_i(k+1)\right\}$$

$$\leqslant P\left\{\left|\frac{1}{k}\sum_{j\in N_i}\varepsilon_{ij}(k)\right| \geqslant h_i(k+1) - \left|z_i(k) - \frac{1}{k}\sum_{j\in N_i}(z_i(k) - z_j(k))\right|\right\}$$

Since

$$z_i(k) - \frac{1}{k}\sum_{j\in N_i}(z_i(k) - z_j(k)) = \left(1 - \frac{|N_i|}{k}\right)z_i(k) + \frac{1}{k}\sum_{j\in N_i}z_j(k)$$

where $|N_i|$ is the number of the elements in the set $N_i$.

On the other hand, if $k > |N_i|$, we have

$$\left(1 - \frac{|N_i|}{k}\right)z_i(k) + \frac{1}{k}\sum_{j\in N_i}z_j(k)$$

$$\leqslant \left(1 - \frac{|N_i|}{k}\right)z_i(k) + \frac{|N_i|}{k}\max_{j\in N_i} z_j(k)$$

$$\leqslant \max\{z_i(k), \max_{j\in N_i} z_j(k)\}$$

$$\leqslant \max_{i=1,\cdots,n} z_i(k)$$

and

$$\left(1 - \frac{|N_i|}{k}\right)z_i(k) + \frac{1}{k}\sum_{j\in N_i}z_j(k)$$

$$\geqslant \left(1 - \frac{|N_i|}{k}\right)z_i(k) + \frac{|N_i|}{k}\min_{j\in N_i} z_j(k)$$

$$\geqslant \min\{z_i(k), \max_{j\in N_i} z_j(k)\}$$

$$\geqslant \max_{i=1,\cdots,n} z_i(k)$$

Thus

$$\left| z_i(k) - \frac{1}{k}\sum_{j\in N_i}(z_j(k) - z_j(k)) \right| \leq \max_{i=1,\cdots,n} |z_i(k)| < h_i(k)$$

Hence

$$D(k) \leq P\left\{ \left| \frac{1}{k}\sum_{j\in N_i} e_{ij}(k) \right| \geq h_i(k+1) - h_i(k) \right\}$$

Using Chebyshev's inequality, we have

$$D(k) \leq P\left\{ \left| \frac{1}{k}\sum_{j\in N_i} \varepsilon_{ij}(k) \right| \geq \frac{1}{(k+1)^{1+\gamma}} \right\}$$

$$\leq \frac{|N_i|\sum_{j\in N_i} E(\varepsilon_{ij}(k))^2}{\dfrac{k^2}{(k+1)^{2(1+\gamma)}}}$$

$$= O\left(k^{2\gamma}\sum_{j\in N_i} E(\varepsilon_{ij}(k))^2\right)$$

It is known by Theorem 4.2

$$E(\varepsilon_{ij}(k))^2 = O\left(\frac{1}{L_k}\right) = O\left(\frac{1}{k^\alpha}\right)$$

Then

$$D(k) = O\left(\frac{1}{k^{\alpha-2\gamma}}\right)$$

Together with Eq. (4.28), we can get that

$$ER_{ij}^2(k) = O\left(\frac{1}{k^{\alpha-2\gamma}}\right) + O\left(\frac{1}{k^\alpha}\right)$$

$$= O\left(\frac{1}{k^{\alpha-2\gamma}}\right)$$

**Theorem 4.3** (Weak Consensus) Let holding time $L_k = k^\alpha$ and the parameter $\gamma$ in the state bound satisfy $0 < \gamma < \alpha/2$. The states updated by Eq. (4.24) can achieve weak consensus under Assumptions 4.1 and 4.2, which means

$$\lim_{t\to\infty} E|x_i(t) - x_j(t)|^2 = 0 \quad \text{for all} \quad i, j = 1, \cdots, n, \ i \neq j$$

**Proof** Denote

$$P_i(k) = \frac{1}{2}\sum_{j\in N_i} |z_i(k) - z_j(k)|^2$$

$$P_N(k) = \sum_{i=1}^{n} P_i(k), \quad V(k) = E(P_N(k))$$

Then

$$P_N(k) = \frac{1}{2}\sum_{i=1}^{n}\sum_{j\in N_i} |z_i(k) - z_j(k)|^2 = z^T(k)Lz(k)$$

Together with Eq. (4.24), we have

$$\begin{aligned}P_N(k+1) &= z^T(k+1)Lz(k+1)\\ &= z^T(k)\left(L - \frac{2}{k}L^2 + \frac{1}{k^2}L^3\right)z(k) + \frac{2}{k}z^T(k)\left(I - \frac{1}{k}L\right)\cdot\\ &\quad L(\varepsilon(k) + R(k)) + \frac{1}{k^2}(\varepsilon(k) + R(k))^T L(\varepsilon(k) + R(k))\end{aligned}$$

(4.29)

By Lemma 4.1, there exist $\beta_1$ and $\beta_2$ such that

$$L - \frac{1}{k}L^2 + \frac{1}{k^2}L^3 \leq \left(1 - \frac{2\beta_1}{k} + \frac{\beta_2}{k^2}\right)L$$

Besides, there exists $K = \frac{\beta_2}{\beta_1}$, for any $k > K$, such that

$$1 - \frac{2\beta_1}{k} + \frac{\beta_2}{k^2} \leq 1 - \frac{\beta_1}{k}$$

Then we have

$$\begin{aligned}&z^T(k)\left(L - \frac{2}{k}L^2 + \frac{1}{k^2}L^3\right)z(k)\\ &\leq \left(1 - \frac{\beta_1}{k}\right)z^T(k)Lz(k) = \left(1 - \frac{\beta_1}{k}\right)P_N(k), \text{ as } k > K\end{aligned}$$ (4.30)

By Theorem 4.2 and Lemma 4.2,

$$\begin{aligned}&E[(\varepsilon(k) + R(k))^T L(\varepsilon(k) + R(k))]\\ &\leq \lambda_{\max}(L)E[(\varepsilon(k) + R(k))^T(\varepsilon(k) + R(k))]\\ &\leq 2\lambda_{\max}(L)E[(\varepsilon(k)^T\varepsilon(k) + R^T(k)R(k))]\\ &= O\left(\frac{1}{k^{\alpha-2\gamma}}\right)\end{aligned}$$ (4.31)

## 4.2 Two-Time-Scale Consensus

Together with Remark 4.2, we can get that

$$E\left|z^T(k)\left(I - \frac{1}{k}L\right)L(\varepsilon(k) + R(k))\right|$$
$$\leq O\left(\sqrt{E[(\varepsilon(k) + R(k))^T(\varepsilon(k) + R(k))]}\right) \quad (4.32)$$
$$= O\left(\frac{1}{k^{\alpha/2-\gamma}}\right)$$

Denote $V(k) = EP_N(k)$. Taking expectation on Eq. (4.29), by Eq. (4.30), Eq. (4.31) and Eq. (4.32), we have

$$V(k+1) \leq \left(1 - \frac{\beta_1}{k}\right)V(k) + O\left(\frac{1}{k^{1+\alpha/2-\gamma}}\right), \quad \text{as } k > K$$

Thus

$$V(k+1) \leq \prod_{i=K}^{k}\left(1 - \frac{\beta_1}{i}\right)V(k) + O\left(\sum_{l=K}^{k-1}\prod_{i=l+1}^{k}\left(1 - \frac{\beta_1}{i}\right)\frac{1}{l^{1+\alpha/2-\gamma}}\right)$$

By Theorem 2.10, we have

$$V(k) = \begin{cases} O\left(\dfrac{1}{k^{\alpha/2-\gamma}}\right) & \alpha/2 - \gamma < \beta_1 \\ O\left(\dfrac{\ln k}{k^{\beta_1}}\right) & \alpha/2 - \gamma = \beta_1 \\ O\left(\dfrac{1}{k^{\beta_1}}\right) & \alpha/2 - \gamma > \beta_1 \end{cases} \quad (4.33)$$

Since $\dfrac{\alpha}{2} - \gamma > 0$ and $\beta_1 > 0$, then

$$V(k) \to 0, \quad \text{as} \quad k \to \infty$$

which implies

$$\lim_{k \to \infty} E|z_i(k) - z_j(k)|^2 = 0, \quad \text{for all} \quad i = 1, \cdots, n, \, j \in N_i$$

Due to the connectivity of $G$, we have

$$\lim_{k \to \infty} E|z_i(k) - z_j(k)|^2 = 0, \quad \text{for all} \quad i, j = 1, \cdots, n, \, i \neq j \quad (4.34)$$

For any $t \in Z^+$, there exists a $k_1$ such that $t \in [t_{k_1}, t_{k_1+1})$, thus

$$\lim_{t \to \infty} E|x_i(t) - x_j(t)|^2 = \lim_{k_1 \to \infty} E|z_i(k_1) - z_j(k_1)|^2$$

By Eq. (4.34), we have

$$\lim_{t \to \infty} E|x_i(t) - x_j(t)|^2 = 0, \quad \text{for all} \quad i, j = 1, \cdots, n, \, i \neq j \quad (4.35)$$

Hence, we can get the theorem.

**Theorem 4.4** (Mean Square Consensus)  Let holding time $L_k = k^\alpha$ and the parameter $\gamma$ in the state bound satisfy $0 < \gamma < \alpha/2$. The states updated by Eq. (4.24) can achieve mean square consensus under Assumptions 4.1 and 4.2, which means there exists a random variable $x^*$ such that

$$\lim_{t \to \infty} E(x_i(t) - x^*)^2 = 0, \quad \text{for all} \quad i = 1, 2, \cdots, n$$

**Proof**  Since $\mathbb{1}^T L = 0$, we have

$$\mathbb{1}^T z(k+1) = \mathbb{1}^T z(1) + \sum_{i=1}^{k} \frac{1}{i} \mathbb{1}^T (\varepsilon(i) + R(i))$$

Denote $\tilde{R}(k) = (\varepsilon(k) + R(k))$, then

$$E(\mathbb{1}^T z(k+m) - \mathbb{1}^T z(k))^2$$

$$= E\left(\sum_{i=k+1}^{k+m} \frac{1}{i} \mathbb{1}^T \tilde{R}(i)\right)^2$$

$$= E\left(\sum_{i=k+1}^{k+m} \sum_{j=k+1}^{k+m} \frac{1}{i} \frac{1}{j} \mathbb{1}^T \tilde{R}(i) \mathbb{1}^T \tilde{R}(j)\right)$$

$$\leq \sum_{i=k+1}^{k+m} \sum_{j=k+1}^{k+m} \frac{1}{i} \frac{1}{j} \sqrt{E(\mathbb{1}^T \tilde{R}(i))^2 E(\mathbb{1}^T \tilde{R}(j))^2}$$

It is known from Eq. (4.31) that

$$E(\tilde{R}^T(k) \tilde{R}(k)) = O\left(\frac{1}{k^{\alpha - 2\gamma}}\right), \quad \text{as} \quad k \to \infty$$

Thus

$$E(\mathbb{1}^T z(k+m) - \mathbb{1}^T z(k))^2$$

$$\leq \sum_{i=k+1}^{k+m} \frac{1}{i^{1+\alpha/2-\gamma}} \sum_{j=k+1}^{k+m} \frac{1}{j^{1+\alpha/2-\gamma}} \to 0, \quad \text{as} \quad k \to \infty$$

According to Theorem 2.9, there exists a random variable $S_\infty \triangleq \mathbb{1}^T z(1) + \sum_{i=1}^{\infty} \frac{1}{i} \mathbb{1}^T \tilde{R}(i)$ such that

$$\lim_{k \to \infty} E(\mathbb{1}^T z(k) - S_\infty)^2 = 0$$

On one hand, we know

$$nz_1(k) = \mathbb{1}_n^T z(k) + (z_1(k) - z_2(k)) + \cdots + (z_1(k) - z_n(k))$$

On the other hand, we have by Theorem 4.3

$$\lim_{k\to\infty} E(z_1(k) - z_j(k))^2 = 0, \ j \neq 1, \quad \text{as} \quad k \to \infty$$

Hence,

$$\lim_{k\to\infty} E(nz_1(k) - \mathbb{1}^T z(\infty))^2 = 0, \quad \text{as} \quad k \to \infty$$

which implies

$$\lim_{k\to\infty} E(z_1(k) - x^*) = 0$$

where $x^* = \dfrac{S_\infty}{n}$.

Analogously, we have

$$\lim_{k\to\infty} E(z_j(k) - x^*) = 0, \ j = 2, \cdots, n, \quad \text{as} \quad k \to \infty$$

For any $t \in Z^+$, there exists a $k_1$ such that $t \in [t_{k_1}, t_{k_1+1})$ thus for all $i = 1, 2, \cdots, n$,

$$\lim_{t\to\infty} E(x_i(t) - x^*)^2 = \lim_{k_1\to\infty} E(z_i(k_1) - x^*)^2 = 0$$

### 4.2.2.2 Convergence Speed

From Theorem 4.4, we have the states $x_i(t)$ converge to $x^*$ in mean square sense. But what's the convergence speed? The following theorem will give the answer.

**Theorem 4.5** (Convergence Speed) Under Assumptions 4.1 and 4.2, the mean square convergence speed of the state $x(t)$, updated by Eq. (4.24) with the holding time $L_k = k^\alpha$, is as follows:

$$E(x_i(t) - x^*)^2 = \begin{cases} O\left(\dfrac{1}{t^{(\alpha-2\gamma)/(2(1+\alpha))}}\right) & \alpha < 2(\beta_1 + \gamma) \\ O\left(\dfrac{\log t}{t^{\beta_1/(\alpha+1)}}\right) & \alpha = 2(\beta_1 + \gamma) \\ O\left(\dfrac{1}{t^{\beta_1/(1+\alpha)}}\right) & \alpha > 2(\beta_1 + \gamma) \end{cases} \quad (4.36)$$

where $\gamma$ is the parameter in the state bound $h_i(k)$ in Eq. (4.20), $\beta_1 = \hat{\lambda}_2 \lambda_n^{-1}$ is given in Lemma 4.1, $\lambda_n$ and $\hat{\lambda}_2$ are respectively the biggest eigenvalue of Laplacian Matrix $L$ and the smallest nonzero eigenvalue of $L^2$.

**Proof** Firstly, let's see the skipping state $z_i(k)$ for any $i=1,\cdots,n$,

$$E(z_i(k) - x^*)^2$$
$$= \frac{1}{n^2} E(nz_i(k) - \mathbb{1}^T z(k) + \mathbb{1}^T z(k) - S_\infty)^2$$
$$= \frac{2}{n^2} E\left(\sum_{j=1}^n (z_i(k) - z_j(k))\right)^2 + \frac{2}{n^2} E(\mathbb{1}^T z(k) - S_\infty)^2$$
$$\leq \frac{2}{n} \max_i E(z_i(k) - z_j(k))^2 + \frac{2}{n^2} E\left(\sum_{i=k+1}^\infty \frac{1}{i} \mathbb{1}^T \tilde{R}(i)\right)^2 \quad (4.37)$$

On one hand, since $G$ is connected, for any two nodes $i$ and $j$, there exist nodes $n_1, n_2, \cdots n_r$ such that

$$i \in N_{n_1},\ \sim n_1 \in N_{n_2},\ \sim n_2 \in N_{n_3},\ \sim \cdots,\ \sim n_{r-1} \in N_{n_r},\ \sim n_r \in N_j$$

Then, we have

$$E(z_i(k) - z_j(k))^2$$
$$= E[(z_i(k) - z_{n_1}(k)) + (z_{n_1}(k) - z_{n_2}(k)) + \cdots + (z_{n_r}(k) - z_j(k))]^2$$
$$\leq (n_r + 1)[E(z_i(k) - z_{n_1}(k))^2 + \cdots + E(z_{n_r}(k) - z_j(k))^2]$$
$$\leq \frac{n_r + 1}{2} \sum_{i=1}^n \sum_{j \in N_i} E(z_i(k) - z_j(k))^2 = (n_r + 1)V(k)$$

Thus, for any $i,j=1,\cdots,n$, we have by Eq. (4.33)

$$E(z_i(k) - z_j(k))^2 = \begin{cases} O\left(\dfrac{1}{k^{\alpha/2-\gamma}}\right) & \alpha < 2(\beta_1 + \gamma) \\ O\left(\dfrac{\ln k}{k^{\beta_1}}\right) & \alpha = 2(\beta_1 + \gamma) \\ O\left(\dfrac{1}{k^{\beta_1}}\right) & \alpha > 2(\beta_1 + \gamma) \end{cases} \quad (4.38)$$

On the other hand

$$E\left(\sum_{i=k+1}^\infty \frac{1}{i} \mathbb{1}^T \tilde{R}(i)\right)^2$$
$$= E\left(\sum_{i=k+1}^\infty \frac{1}{i} \mathbb{1}^T \tilde{R}(i) \sum_{j=k+1}^\infty \frac{1}{j} \mathbb{1}^T \tilde{R}(j)\right)$$
$$\leq \sum_{i=k+1}^\infty \sum_{j=k+1}^\infty \frac{1}{i} \frac{1}{j} \sqrt{E(\mathbb{1}^T \tilde{R}(i))^2 E(\mathbb{1}^T \tilde{R}(j))^2}$$
$$\leq \sum_{i=k+1}^\infty \frac{1}{i^{1+\alpha/2-\gamma}} \sum_{j=k+1}^\infty \frac{1}{j^{1+\alpha/2-\gamma}} = O\left(\frac{1}{k^{\alpha-2\gamma}}\right) \quad (4.39)$$

## 4.2 Two-Time-Scale Consensus

By Eq. (4.37), Eq. (4.38) and Eq. (4.39), we have

$$E(z_i(k) - x^*)^2 = \begin{cases} O\left(\dfrac{1}{k^{\alpha/2-\gamma}}\right) & \alpha < 2(\beta_1 + \gamma) \\ O\left(\dfrac{\ln k}{k^{\beta_1}}\right) & \alpha = 2(\beta_1 + \gamma) \\ O\left(\dfrac{1}{k^{\beta_1}}\right) & \alpha > 2(\beta_1 + \gamma) \end{cases} \quad (4.40)$$

Secondly, let's see $E(x_i(t) - x^*)^2$. Since $z_i(k) = x_i(t)$, $t = t_k, \cdots, t_{k+1} - 1$, then we have

$$E(z_i(k) - x^*)^2 = E(x_i(t) - x^*)^2, \; t \in [t_k, t_{k+1}), \; k = 1, \cdots, n \quad (4.41)$$

For $t \in [t_k, t_{k+1})$, $k = 1, 2, \cdots$, we obtain

$$t \geq t_k = t_{k-1} + L_{k-1} = \sum_{i=1}^{k-1} L_i = \sum_{l=1}^{k-1} l^\alpha \geq (k-1)^{\alpha+1} - 1$$

$$t \leq t_{k+1} = \sum_{l=1}^{k} l^\alpha \leq (k+1)^{\alpha+1} - 1$$

Thus, we can get

$$(t+1)^{\frac{1}{\alpha+1}} - 1 \leq k \leq (t+1)^{\frac{1}{\alpha+1}} + 1$$

which implies

$$k \to \infty \Leftrightarrow t \to \infty$$

and $k$ is equivalent to $t^{\frac{1}{\alpha+1}}$. Replacing $k$ by $t^{\frac{1}{\alpha+1}}$ in Eq. (4.40), together with Eq. (4.41), we have for $t \in [t_k, t_{k+1})$

$$E(x_i(t) - x^*)^2 = \begin{cases} O\left(\dfrac{1}{t^{(\alpha-2\gamma)/(2(1+\alpha))}}\right) & \alpha < 2(\beta_1 + \gamma) \\ O\left(\dfrac{\log t}{t^{\beta_1/(\alpha+1)}}\right) & \alpha = 2(\beta_1 + \gamma) \\ O\left(\dfrac{1}{t^{\beta_1/(1+\alpha)}}\right) & \alpha > 2(\beta_1 + \gamma) \end{cases}$$

Since the above equation is true for all $k = 1, 2, \cdots$, it is true for all $t = 1, 2, \cdots$. Hence, Theorem 4.5 is proved.

**Remark 4.3** For the skipping state $x(t_k)$, $k = 1, 2, \cdots$, the convergence speed is given in expression (4.40). If $\alpha < 2\beta_1$, we can see that the larger $\alpha$ is, the higher the convergence speed will be. If $\alpha > 2\beta_1$, the convergence speed will be unchanged even if $\alpha$ becomes bigger. When the communication network $G$ of the system is given, the communication network $G$ of the system is given, the matrix $L$ will be given and $\beta_1$ will be found. We can choose a larger $\alpha$ such that $\alpha > 2\beta_1$, then the state $z(k)$ will converge to $x^* \mathbb{1}_n$ with the convergence speed $O\left(\dfrac{1}{k^{\beta_1}}\right)$, which is consistent with the fact that the more time each node spends to estimate its neighbors' states, the more accurate the estimate, the faster the state $z(k)$ converges to $x^* \mathbb{1}_n$.

**Remark 4.4** For the state $x_i(t)$, $i = 1, 2, \cdots, n$, the convergence speed is given in expression (4.36). If $\alpha < 2(\beta_1 + \gamma)$, then

$$E(x_i(t) - x^*)^2 = O\left(\dfrac{1}{t^{(\alpha - 2\gamma)/(2(1+\alpha))}}\right)$$

which implies that the large $\alpha$ is, the slower the states converge, which is due to long time the states keeping unchanged. So, the parameter $\alpha$ should not be too big or too small in order to get a higher convergence speed.

**Remark 4.5** The convergence speed given in Eq. (4.36) is slower than that of $O\left(\dfrac{1}{\sqrt{t}}\right)$. However, from the proof of Theorem 4.5, we can see that the convergence speed given in Eq. (4.36) is achieved by $\leqslant$ relation, which implies the true value of $E(x_i(t) - x^*)^2$ may be smaller than that in Eq. (4.36). In another word, the true convergence speed may be faster than that in Eq. (4.36).

### 4.2.3 The Consensus Protocol

Summarizing Section 4.2.1 and Section 4.2.2, the whole Consensus Protocol can be given as follows:

(1) Initiation: Let $k = 1$ and $t_1 = 0$;

(2) Observations:

Let $t_{k+1} = t_k + L_k$, where $L_k = k^\alpha$, $\alpha > 0$. In the time interval $[t_k, t_{k+1})$, the states of all the nodes will be unchanged, which means

$$x_i(t_k) = x_i(t_k + 1) = \cdots = x_i(t_{k+1} - 1) \triangleq z_i(k)$$

By Eq. (4.3), each node $i$ can get the $L_k$ binary-valued measurements in time interval $[t_k, t_{k+1})$ from its neighbors:

$$\begin{cases} y_{ij}(l) = x_j(t_k) + d_{ij}(l) \\ s_{ij}(l) = I_{\{y_{ij}(l) \leqslant C\}} \end{cases} \quad j \in N_i, \ l = t_k, \cdots, t_{k+1} - 1$$

(3) Estimation: For each node $i$, the frequency that it gets the binary valued measurement 1 from neighbor $j$ for the holding time $L_k$ can be given by

$$\varphi_{ij}(L_k) = \frac{1}{L_k} \sum_{l=t_k}^{t_k+L_k-1} s_{ij}(l), \quad j \in N_i, \quad i = 1, \cdots, n$$

Since $F(\cdot)$ is not invertible at 0 and 1, we modify $\varphi_{ij}(L_k)$ to $\xi_{ij}(L_k)$ to avoid these points.

Let

$$\xi_{ij}(L_k) = \begin{cases} \frac{1}{2}, & \text{if } \varphi_{ij}(L_k) = 0 \text{ or } 1 \\ \varphi_{ij}(L_k), & \text{if } 0 < \varphi_{ij}(L_k) < 1 \end{cases}$$

Then, each node $i$ estimates its neighbor $j$'s state by

$$\hat{z}_{ij}(L_k) = C - F^{-1}(\xi_{ij}(L_k)), \quad j \in N_i, \quad i = 1, \cdots, n \tag{4.42}$$

(4) Control: Based on estimate $\hat{z}_{ij}(L_k)$, node $i$ designs the control $u_i(k)$ at time by the averaging rule with the step size $1/k$:

$$u_i(k) = -\frac{1}{k} \sum_{j \in N_i} (z_i(k) - \hat{z}_{ij}(L_k)) \cdot I_{\{|z_i(k) - \frac{1}{k}\sum_{j \in N_i}(z_i(k) - \hat{z}_{ij}(L_k))| < h_i(k+1)\}} \tag{4.43}$$

where $u_i(k) = u_i(t_{k+1} - 1)$, $\hat{z}_{ij}(L_k)$ is the estimate of neighbor $j$'s state $z_j(k)$, $h_i(k+1) = M_i + \sum_{h=1}^{k+1} \frac{1}{h^{1+\gamma}}$ is the bound of agent $i$'s state with $M_i > |x_i(0)|$, $0 < \gamma < \alpha/2$. By Eq. (4.1), the state of node $i(i=1, \cdots, n)$ will be updated at time $t_{k+1}$ by.

$$z_i(k+1) = z_i(k) - \frac{1}{k} \sum_{j \in N_i} (z_i(k) - \hat{z}_{ij}(L_k)) I_{\{|z_i(k) - \frac{1}{k}\sum_{j \in N_i}(z_i(k) - \hat{z}_{ij}(L_k))| < h_i(k+1)\}}$$

(5) Let $k = k+1$, go back to step (2).

### 4.2.4 Numerical Simulation

Numerical simulations will be given to illustrate the theoretical results obtained in the previous sections. Here we consider the network computing problem as Fig. 4.2. The goal is to make the work the workload of different computers (agents) consensus.

The initial state of the system is set to be $x(0) = [0, 1, 3, 8, 13]^T$. For a holding time $L_k = k^\alpha$ ($\alpha = 0.8, 2, 3, 4$), respectively in the simulations, each agent can get binary-valued measurements from its neighbors for the holding time $L_k$ by Eq. (4.4), where the noise $d_{ij}(l)$ is normal distribution $N(0, 5)$ and the threshold $C = 3$. Based on the measurements $s_{ij}(l)$, each agent estimates its neighbors' states by Eq. (4.42). Then it can design the control by Eq. (4.43), where $M_i = 25$ for all $i = 1, \cdots, n$ and $\gamma = \alpha/3$. In this example, $\tilde{u}_1(k) = -\frac{1}{k}[z_1(k) - \hat{z}_{12}(L_k) + z_1(k) - \hat{z}_{14}(L_k) + z_1(k) - $

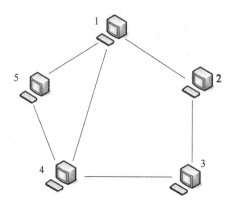

Fig. 4.2  Network computing with five computers

$\hat{z}_{15}(L_k)]$ and $\tilde{u}_2(k) = -\dfrac{1}{k}[z_2(k) - \hat{z}_{21}(L_k) + z_2(k) - \hat{z}_{23}(L_k)]$, etc. Finally the states of the five nodes are updated by Eq. (4.22).

The results can be shown in the following figures. Fig. 4.3 and Fig. 4.4 are the trajectories of $z_i(k)$, $i=1, \cdots, 5$ and $x_i(k)$, $i=1, \cdots, 5$, respectively. Fig. 4.5 is the variance of $x(t)$ with different $\alpha$. From Fig. 4.3, we can see the trajectories of the five nodes converge to a same state with different holding time $L_k$. Comparing 4.3(a) ~ (d), we can see that $z_i(k)$, $i=1, \cdots, 5$, converge faster as $\alpha$ becomes larger, which is consistent with Remark 4.3.

From Fig. 4.4, we can see the states of all the five nodes converge to a same state for any $\alpha > 0$. The variances converge to zero, which is shown in Fig. 4.5. But $x_i(t)$, $i=1, \cdots, 5$, doesn't always converge faster as $\alpha$ becomes larger. When $\alpha$ changes from 0.8 to 2, $x_i(t)$ converge faster, which can be seen form Fig. 4.4(a)、(b). But when

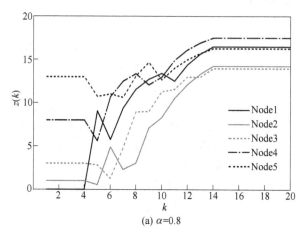

(a) $\alpha=0.8$

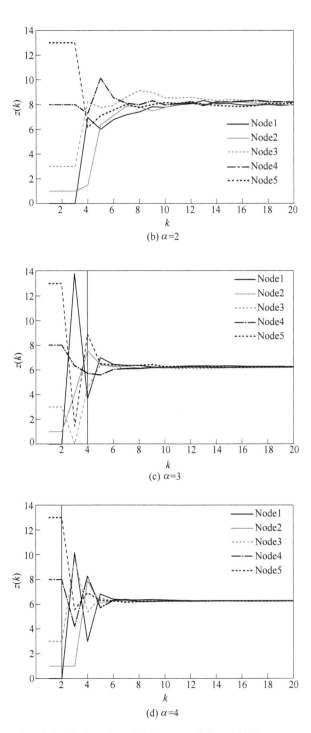

Fig. 4.3 Trajectories of skip states $z(k)$ with different $\alpha$

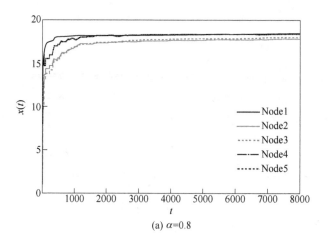

(a) $\alpha=0.8$

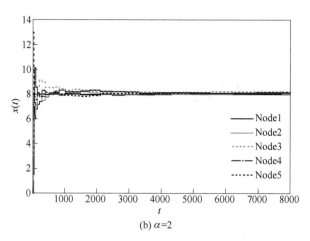

(b) $\alpha=2$

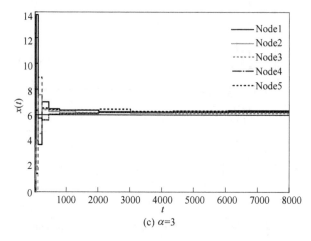

(c) $\alpha=3$

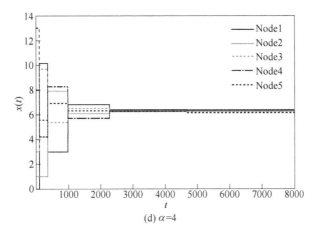

(d) $\alpha=4$

Fig. 4.4  Trajectories of states $x(t)$ with different $\alpha$

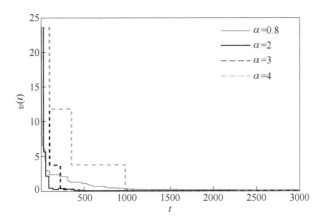

Fig. 4.5  Variance of $x(t)$ with different $\alpha$

$\alpha=4$, the states $x_i(t)$, $i=1, \cdots, 5$, shown in Fig. 4.4(d) converge slower than that in Fig. 4.4(c). In another word, the convergence speed becomes faster and then slower with the increase of $\alpha$, which can be seen from Fig. 4.5, too. This result is consistent with Theorem 4.5.

In the simulations, we can conclude that the Consensus Protocol proposed in this chapter can deal with the consensus problems of the multi-agent systems with binary-valued communications.

## 4.3  Recursive Projection Consensus

In this section, the objective is still to design the control $u_i(t)$, $i=1, \cdots, n$, by binary-valued communications (4.3), to make the agents in system (4.1) achieve consensus. Another algorithm——recursive projection consensus algorithm is proposed to complete the task.

### 4.3.1 Consensus Algorithm

Applying the recursive projection algorithm Eq. (3.29) to the sate estimation of communication system (4.3), each agent can get the estimates of the neighbors' states. Replacing the neighbors' states with the estimates, the consensus control with step size $1/(t+1)$ is given by the average rule. Specifically, the algorithm is summarized by Consensus Algorithm with the following 5 steps.

(1) Initiation. Let the initial state and the initial estimate of agent $i$ be respectively as follows:

$$x_i(1) = x_i^0 \quad \hat{x}_{ij}(0) = \hat{x}_{ij}^0$$

for $j \in N_i$, $i = 1, \cdots, n$, where $|x_i^0| \leq M$, $|\hat{x}_{ij}^0| \leq M$, $M > 0$ is a given constant.

(2) Observation. Each agent $i$ gets the binary-valued observations from its neighbors:

$$\begin{cases} y_{ij}(t) = x_j(t) + w_{ij}(t) \\ s_{ij}(t) = I_{\{y_{ij}(t) \leq C\}} \end{cases} \quad j \in N_i, \ i = 1, \cdots, n$$

which is the same as Eq. (4.3).

(3) Estimation. Based on the recursive projection algorithm Eq. (3.29) to state estimation, each agent $i$ estimates its neighbors' states by

$$\hat{x}_{ij}(t) = \prod_M \left\{ \hat{x}_{ij}(t-1) + \frac{\beta}{t}(F(C - \hat{x}_{ij}(t-1)) - s_{ij}(t)) \right\} \quad (4.44)$$

where $\prod_M(\cdot)$ is a projection operator, which is defined as follows:

$$\prod_M(x) = \arg\min_{|m| \leq M} |x - m| = \begin{cases} -M & \text{if } x < -M \\ x & \text{if } |x| \leq M \\ M & \text{if } x > M \end{cases}$$

(4) Control. Based on the estimates, each agent designs the control with a decreasing consensus gain $1/(t+1)$:

$$u_i(t) = -\frac{1}{t+1} \sum_{j \in N_i} (x_i(t) - \hat{x}_{ij}(t)), \ i = 1, \cdots, n$$

By the control, the state of agent $i$ is updated by

## 4.3 Recursive Projection Consensus

$$x_i(t+1) = x_i(t) - \frac{1}{t+1}\sum_{j\in N_i}(x_i(t) - \hat{x}_{ij}(t)) \tag{4.45}$$

(5) Let $t = t+1$, go back to (2).

**Remark 4.6** The form of state estimation Eq. (4.44) is similar to the form of the algorithm Eq. (3.29). But state estimation Eq. (4.44) is not parameter identification as algorithm Eq. (3.29) because the states are not constant, which are updated by the control. So the analysis of state estimation Eq. (4.44) will involve the state updating, which is different from the analysis of algorithm Eq. (3.29).

**Remark 4.7** Due to the definition of the projection operator, we have

$$|\hat{x}_{ij}(t)| \le M, \ j \in N_i, \ i = 1, \cdots, n$$

**Proposition 4.1** The state of agent $i$ satisfies

$$|x_i(t)| \le M$$

if $t \ge d_i$. Moreover, all the states satisfy

$$|x_i(t)| \le M, \ \forall i = 1, \cdots, n$$

if $t \ge d_*$, where $d_*$ is the maximum degree of the nodes in the system network $G$, i.e. $d_* = \max\{d_i, \ i = 1, \cdots, n\}$.

**Proof** By the state updating Eq. (4.45), we have

$$|x_i(t)| = \left|x_i(t-1) - \frac{1}{t}\sum_{j\in N_i}(x_i(t-1) - \hat{x}_{ij}(t-1))\right|$$

$$= \left|\left(1 - \frac{d_i}{t}\right)x_i(t-1) + \frac{1}{t}\sum_{j\in N_i}\hat{x}_{ij}(t-1)\right|$$

$$\le \left|1 - \frac{d_i}{t}\right||x_i(t-1)| + \frac{d_i}{t}M$$

If $t = d_i$, we can get

$$|x_i(d_i)| \le M$$

Assume the inequality $|x_i(t)| \le M$ holds when $t = d_i + m$ for any $m \in N$. Then, when $t = d_i + m + 1$, we have

$$|x_i(t)| \le \left(1 - \frac{d_i}{d_i + m + 1}\right)|x_i(d_i + m)| + \frac{d_i}{d_i + m + 1}M$$

$$\le \left(1 - \frac{d_i}{d_i + m + 1}\right)\max\{|x_i(d_i + m)|, M\} +$$

$$\frac{d_i}{d_i + m + 1}\max\{|x_i(d_i + m)|, M\}$$

$$= \max\{|x_i(d_i + m)|, M\} \le M$$

By mathematical induction, we can get

$$|x_i(t)| \leq M, \text{ if } t \geq d_i$$

Moreover, if $t \geq d_* = \max\{d_i, i = 1, \cdots, n\}$, then

$$|x_i(t)| \leq M, \forall i = 1, \cdots, n$$

The proposition is proved.

By state updating Eq. (4.45), we have

$$x_i(t+1) = x_i(t) - \frac{1}{t+1}\sum_{j \in N_i}(x_i(t) - x_j(t)) + \frac{1}{t+1}\sum_{j \in N_i}(\hat{x}_{ij}(t) - x_j(t))$$

Denote $x(t) = (x_1(t), \cdots, x_n(t))^T$, $\varepsilon_{ij}(t) = \hat{x}_{ij}(t) - x_j(t)$. Putting $\varepsilon_{ij}(t)$, $j \in N_i$, $i=1, \cdots, n$ in a given order yields the error vector $\varepsilon(t)$. Without loss of generality, let

$$\varepsilon(t) = (\varepsilon_{1r_1}(t), \cdots, \varepsilon_{1r_{d_1}}(t), \varepsilon_{2r_{d_1+1}}(t), \cdots, \varepsilon_{2r_{d_1+d_2}}(t), \cdots,$$

$$\varepsilon_{nr_{d_1+\cdots+d_{n-1}+1}}(t), \cdots, \varepsilon_{nr_{d_1+\cdots+d_n}}(t))^T \quad (4.46)$$

where $r_1, \cdots, r_{d_1} \in N_1$, $r_{d_1+1}, \cdots, r_{d_1+d_2} \in N_2$, $\cdots, r_{d_1+\cdots+d_{n-1}+1}, \cdots, r_{d_1+\cdots+d_n} \in N_n$. For any $\varepsilon_{ij}(t)$, $j \in N_i$, $i=1, \cdots, n$, we can define $n$ dimensional vectors $p_{ij}$ and $q_{ij}$ respectively to label the start point and the end point as follows:

$$p_{ij} = (0, \cdots, 0, \underset{\text{ith position}}{1}, 0, \cdots, 0)^T$$

$$q_{ij} = (0, \cdots, 0, \underset{\text{jth position}}{1}, 0, \cdots, 0)^T$$

for $j \in N_i$, $i=1, \cdots, n$. Putting $\{p_{ij}, j \in N_i, i=1, \cdots, n\}$ and $\{q_{ij}, j \in N_i, i=1, \cdots, n\}$ in the same order as $\varepsilon(t)$ in Eq. (4.46), we can get matrixes $P$ and $Q$ as follows:

$$P = [p_{1r_1}, \cdots, p_{1r_{d_1}}, p_{2r_{d_1+1}}, \cdots, p_{2r_{d_1+d_2}}, \cdots,$$

$$p_{nr_{d_1+\cdots+d_{n-1}+1}}, \cdots, p_{nr_{d_1+\cdots+d_n}}]$$

$$= \begin{bmatrix} \underbrace{11\cdots1}_{d_1} & & & \\ & \underbrace{11\cdots1}_{d_2} & & 0 \\ & & \ddots & \\ 0 & & & \underbrace{11\cdots1}_{d_n} \end{bmatrix}_{n \times (d_1+\cdots+d_n)} \quad (4.47)$$

## 4.3 Recursive Projection Consensus

$$Q = \begin{bmatrix} q_{1r_1}^T \\ \vdots \\ q_{1r_{d_1}}^T \\ q_{2r_{d_1+1}}^T \\ \vdots \\ q_{2r_{d_1+d_2}}^T \\ \vdots \\ q_{nr_{d_1+d_{n-1}+1}}^T \\ \vdots \\ q_{nr_{d_1+\cdots+d_n}}^T \end{bmatrix}_{(d_1+\cdots+d_n)\times n} \tag{4.48}$$

Then the state updating can be written in a vector form:

$$x(t+1) = \left(I - \frac{L}{t+1}\right)x(t) + \frac{P}{t+1}\varepsilon(t) \tag{4.49}$$

where $L$ is the Laplacian Matrix for the network $G$.

**Example 4.1** Consider the network with 3 agents as Fig. 4.6. By definition Eq. (4.46) Eq. (4.47) and Eq. (4.48) we can get the error vector $\varepsilon(t)$, the corresponding matrixes $P$, $Q$ and the Laplacian Matrix $L$ as follows:

$$\varepsilon(t) = (\varepsilon_{12}(t), \varepsilon_{13}(t), \varepsilon_{21}(t), \varepsilon_{23}(t), \varepsilon_{31}(t), \varepsilon_{32}(t))^T$$

$$P = \begin{bmatrix} 1 & 1 & 0 & 0 & 0 & 0 \\ 0 & 0 & 1 & 1 & 0 & 0 \\ 0 & 0 & 0 & 0 & 1 & 1 \end{bmatrix}, \quad Q = \begin{bmatrix} 0 & 1 & 0 \\ 0 & 0 & 1 \\ 1 & 0 & 0 \\ 0 & 0 & 1 \\ 1 & 0 & 0 \\ 0 & 1 & 0 \end{bmatrix} \tag{4.50}$$

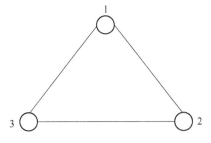

Fig. 4.6 Network computing with 3 nodes

$$L = \begin{bmatrix} 2 & -1 & -1 \\ -1 & 2 & -1 \\ -1 & -1 & 2 \end{bmatrix} \quad (4.51)$$

### 4.3.2 Main Results

In this subsection, we will discuss the properties of Consensus Algorithm in Section 4.3.1. Here, we will introduce the following lemmas firstly.

**Lemma 4.3** For an undirected network with $n$ nodes, matrices $P$ and $Q$ are respectively defined as Eq. (4.47) and Eq. (4.48). Then matrices $P^T P$ and $QQ^T$ have the same eigenvalues as follows

$$\tilde{\lambda}_1 = d_1, \ \tilde{\lambda}_2 = d_2, \ \cdots, \ \tilde{\lambda}_n = d_n$$
$$\tilde{\lambda}_{n+1} = \cdots = \tilde{\lambda}_{d_1+d_2+\cdots+d_n} = 0$$

where $d_i$, $i = 1, \cdots, n$ are the degree of node $i$.

**Proof** Firstly, let's calculate the eigenvalues of the matrix $P^T P$. By definition Eq. (4.47), we have

$$P^T P = \begin{bmatrix} \mathbb{1}_{d_1} & & & & \\ & \mathbb{1}_{d_2} & & 0 & \\ & & \ddots & & \\ & 0 & & \ddots & \\ & & & & \mathbb{1}_{d_n} \end{bmatrix}$$

where $\mathbb{1}_{d_i}$ is the $d_i \times d_i$ dimensional matrix with all the elements being 1 for $i = 1, \cdots, n$. So the determinant of $(\tilde{\lambda} I_{d_1+\cdots+d_n} - P^T P)$ is

$$|\tilde{\lambda} I_{d_1+\cdots+d_n} - P^T P| = |\tilde{\lambda} I_{d_1} - \mathbb{1}_{d_1}||\tilde{\lambda} I_{d_2} - \mathbb{1}_{d_2}|\cdots|\tilde{\lambda} I_{d_n} - \mathbb{1}_{d_n}|$$

where $I_m (\forall m \in Z^+)$ is the $m \times m$ dimensional identity matrix. Since the solutions of the algebraic equation

$$|\tilde{\lambda} I_n - \mathbb{1}_n| = 0$$

are $\tilde{\lambda}_1 = n$, $\tilde{\lambda}_2 = \tilde{\lambda}_3 = \cdots = \tilde{\lambda}_n = 0$, it follows that the solutions of the algebraic equation

$$|\tilde{\lambda} I_{d_1+\cdots+d_n} - P^T P| = 0$$

are

$$\tilde{\lambda}_1 = d_1, \ \tilde{\lambda}_2 = d_2, \ \cdots, \ \tilde{\lambda}_n = d_n$$
$$\tilde{\lambda}_{n+1} = \cdots = \tilde{\lambda}_{d_1+d_2+\cdots+d_n} = 0$$

which are the eigenvalues of the matrix $P^T P$.

Secondly, let's prove that matrixes $P^TP$ and $QQ^T$ have the same eigenvalues. If we reorder the elements of $\{q_{ij}, j \in N_i, i=1, \cdots, n\}$, we can get the matrix $\tilde{Q}$ as

$$\tilde{Q} = \begin{bmatrix} q_{m_1 1} \\ \vdots \\ q_{m_{d_1} 1} \\ q_{m_{d_1}+1 2} \\ \vdots \\ q_{m_{d_1}+d_2 2} \\ \vdots \\ q_{m_{d_1}+\cdots+d_{n-1}+1 n} \\ \vdots \\ q_{m_{d_1}+\cdots+d_n n} \end{bmatrix} = \begin{bmatrix} 1 & 0 & & \\ \vdots & \vdots & & \\ 1 & 0 & & 0 \\ 0 & 1 & & \\ \vdots & \vdots & & \\ 0 & 1 & & \\ & & \ddots & \\ & & 0 & 1 \\ & & \vdots & \vdots \\ & & & 1 \end{bmatrix} = P^T$$

Due to the definition of $Q$ and $\tilde{Q}$, we can get that $Q$ can be obtained by exchanging the row vectors of $\tilde{Q}$. Hence, there exists a sequence of $(d_1+\cdots+d_n) \times (d_1+\cdots+d_n)$ elementary matrices $I_1, I_2, \cdots, I_s$, which are identity matrixes exchanging a pair of row vectors, such that

$$Q = I_1 I_2 \cdots I_s \tilde{Q} = I_1 I_2 \cdots I_s P^T$$

Consequently,

$$QQ^T = I_1 I_2 \cdots I_s P^T P (I_1 I_2 \cdots I_s)^T$$

Since $I_t (t=1, \cdots, s)$ is obtained by exchanging a pair of row vectors of identity matrix, it follows that

$$I_t I_t^T = I_t^T I_t = I_{d_1+\cdots+d_n}, \quad \forall t = 1, \cdots, s$$

Hence, matrix $QQ^T$ is similar to matrix $P^TP$, which implies the eigenvalues of $P^TP$ and $QQ^T$ are the same.

**Lemma 4.4** ([19], Theorem 5)  Under Assumption 4.1, the Laplacian Matrix $L$ follows

$$x^T L^2 x \geq c_1 x^T L x, \quad x^T L^3 x \leq c_2 x^T L x$$

where $c_1 = \dfrac{\lambda_2^2}{\lambda_n}$, $c_2 = \dfrac{\lambda_n^3}{\lambda_2}$, $\lambda_2$ and $\lambda_n$ are the smallest and the biggest positive eigenvalues of Laplacian Matrix $L$, respectively.

Denote the mean square error of state estimation as

$$R(t) = E(\varepsilon(t)^T \varepsilon(t))$$

where $\varepsilon(t)$ is defined as Eq. (4.46). Denote

$$V(t) = E(x^T(t) L x(t))$$

which characterizes the consensus property of the states.

**Lemma 4.5** Under Assumption 4.2, for the estimation Eq. (4.44) in Consensus Algorithm, there exists a positive constant $\check{A}$ such that the mean square error $R(t)$ satisfies

$$R(t) \leqslant \left(1 - \frac{2\beta f_M - \frac{\lambda_n d_*}{\alpha} - 2d_*}{t}\right) R(t-1) + \frac{\alpha}{t} V(t-1) + \frac{\check{A}}{t^2}$$

$$\forall 0 < \alpha < \infty, \text{ as } t > d_* \qquad (4.52)$$

where $V(t-1)$ is the Lyapunov function at time $t-1$, $f_M = f(|C|+M)$, $\beta$ is a constant parameter in the estimation Eq. (4.44), $\lambda_n$ is the maximum eigenvalue of Laplacian Matrix $L$, $d_*$ is the maximum degree of the nodes in the system network.

**Proof** By the estimation Eq. (4.44), the state updating Eq. (4.45) and Proposition 4.1, if $t > d_*$, we have

$$Ee_{ij}^2(t) = E(\hat{x}_{ij}(t) - x_j(t))^2$$

$$\leqslant E\Big(\hat{x}_{ij}(t-1) - x_j(t) + \frac{\beta}{t}(F(C - \hat{x}_{ij}(t-1)) - s_{ij}(t))\Big)^2$$

$$= E\bigg(\varepsilon_{ij}(t-1) + \frac{\beta}{t}(F(C - \hat{x}_{ij}(t-1)) - s_{ij}(t)) + \qquad (4.53)$$

$$\frac{1}{t}\sum_{p \in N_j}(x_j(t-1) - x_p(t-1) - \varepsilon_{jp}(t-1))\bigg)^2$$

Denote

$$\eta_{ij}(t) = \varepsilon_{ij}(t) + \frac{\beta}{t+1}(F(C - \hat{x}_{ij}(t)) - s_{ij}(t+1)) +$$

$$\frac{1}{t+1}\sum_{p \in N_j}(x_j(t) - x_p(t) - \varepsilon_{jp}(t))$$

Putting $\{\eta_{ij}, j \in N_i, i=1, \cdots, n\}$ in the same order as $\varepsilon(t)$ in Eq. (4.46) yields the vector $\eta(t)$:

$$\eta(t) = \varepsilon(t) + \frac{\beta}{t+1}(\hat{F}(t) - S(t+1)) + \frac{1}{t+1}Q(Lx(t) - P\varepsilon(t))$$

$$(4.54)$$

where the elements of $\hat{F}(t)$ and $S(t)$ are respectively $F(C-\hat{x}_{ij}(t))$ and $s_{ij}(t)$ with the same order as $\varepsilon(t)$, the matrixes $Q$ and $P$ are defined as Eq. (4.48) and Eq. (4.47). By Eq. (4.46) and Eq. (4.53), we can obtain

$$R(t) = Ee(t)^T \varepsilon(t) = \sum_{i=1}^n \sum_{j \in N_i} E\varepsilon_{ij}^2(t)$$

$$\leq \sum_{i=1}^n \sum_{j \in N_i} E\eta_{ij}^2(t-1) = E\eta(t-1)^T \eta(t-1)$$

Also, $\eta(t)$ follows Eq. (4.54), we have consequently

$$R(t) \leq R(t-1) + \frac{2\beta}{t} E[\varepsilon(t-1)^T(\hat{F}(t-1) - S(t))] +$$

$$\frac{2}{t} E[\varepsilon(t-1)^T Q L x(t-1)] -$$

$$\frac{2}{t} E[\varepsilon(t-1)^T Q P \varepsilon(t-1)] + \frac{1}{t^2}$$

$$E[(\beta(\hat{F}(t-1) - S(t)) + Q(Lx(t-1) - P\varepsilon(t-1)))^T$$

$$(\beta(\hat{F}(t-1) - S(t)) + Q(Lx(t-1) - P\varepsilon(t-1)))]$$

Since $F(C-\hat{x}_{ij}(t))$ and $s_{ij}(t)$ are bounded, together with Remark 4.7 and Proposition 4.1, as $t > d_*$, there exists $0 < B_1 < \infty$ such that

$$R(t) \leq R(t-1) + \frac{2\beta}{t} E[\varepsilon(t-1)^T(\hat{F}(t-1) - S(t))] +$$

$$\frac{2}{t} E[\varepsilon(t-1)^T Q L x(t-1)] - \qquad (4.55)$$

$$\frac{2}{t} E[\varepsilon(t-1)^T Q P \varepsilon(t-1)] + \frac{B_1}{t^2}$$

Define $D_t = \sigma\{P_{ij}(1), \cdots, P_{ij}(t), i=1, \cdots, n, j \in N_i\}$, then $\hat{x}_{ij}(t)$, $j \in N_i$, $i=1, \cdots, n$ and $\varepsilon_{ij}(t)$, $j \in N_i$, $i=1, \cdots, n$ are $D_t$ measurable. Similar to Eq. (3.32), we have

$$E(\varepsilon_{ij}(t-1)(F(C-\hat{x}_{ij}(t-1)) - s_{ij}(t)))$$

$$= -E[f(\xi_{ij})\varepsilon_{ij}(t-1)(\hat{x}_{ij}(t-1) - x_j(t))]$$

$$= -E\left\{f(\xi_{ij})\varepsilon_{ij}(t-1)\left[\varepsilon_{ij}(t-1) + \frac{1}{t}\sum_{p \in N_j}(x_j(t-1) - \hat{x}_{jp}(t-1))\right]\right\}$$

where $\xi_{ij} \in (C-\hat{x}_{ij}(t-1), C-x_j(t))$ or $\xi_{ij} \in (C-x_j(t-1), C-\hat{x}_{ij}(t-1))$. By Remark 4.7 and Proposition 4.1, $|\xi_{ij}(t)| \leq |C| + M$. Moreover, $f(\xi_{ij}) \geq f(|C|+M) \triangleq f_M$. On the other hand, $f(\xi_{ij}) \leq f(0)$, together with Proposition 4.1, we can get that

there exists $0 < B_2 < \infty$ such that
$$E(\varepsilon_{ij}(t-1)(F(C-\hat{x}_{ij}(t-1))-s_{ij}(t)))$$
$$\leqslant -f_M E\varepsilon_{ij}(t-1)^2 + \frac{B_2}{t}, \text{ as } t > d_*$$

Thus, the second item on the right side of inequality (4.55) will be
$$\frac{2\beta}{t}E[\varepsilon(t-1)^T(\hat{F}(t-1)-S(t))]$$
$$=\frac{2\beta}{t}\sum_{i=1}^{n}\sum_{j\in N_i}E[\varepsilon_{ij}(t-1)(F(C-\hat{x}_{ij}(t-1))-s_{ij}(t))]$$
$$\leqslant -\frac{2\beta f_M}{t}R(t-1)+\frac{2\beta n N_i B_2}{t^2} \tag{4.56}$$

Noticing that there exists $\tilde{L}$ such that
$$L = \tilde{L}^T\tilde{L} \tag{4.57}$$

The third item on the right side of inequality (4.55) is
$$\frac{2}{t}E[\varepsilon(t-1)^T QLx(t-1)]$$
$$\leqslant \frac{2}{t}E[\varepsilon(t-1)^T Q\tilde{L}^T\tilde{L}x(t-1)]$$
$$\leqslant \frac{2}{t}\sqrt{E[\varepsilon(t-1)^T QLQ^T\varepsilon(t-1)]E[x(t-1)^T Lx(t-1)]}$$

Denote $\lambda_Q = \lambda_{\max}(QQ^T)$, we have
$$\frac{2}{t}E[\varepsilon(t-1)^T QLx(t-1)]$$
$$\leqslant \frac{2}{t}\sqrt{\frac{\lambda_n \lambda_Q}{\alpha}R(t-1)\alpha V(t-1)}$$
$$\leqslant \frac{1}{t}\left(\frac{\lambda_n \lambda_Q}{\alpha}R(t-1)+\alpha V(t-1)\right) \tag{4.58}$$

where $\alpha > 0$. Moreover, the fourth item on the right side of inequality (4.55) is
$$-\frac{2}{t}E[\varepsilon(t-1)^T QP\varepsilon(t-1)]$$
$$\leqslant \frac{2}{t}\sqrt{E[\varepsilon(t-1)^T QQ^T\varepsilon(t-1)]}\sqrt{E[\varepsilon(t-1)^T P^T P\varepsilon(t-1)]} \tag{4.59}$$
$$\leqslant \frac{2\sqrt{\lambda_Q \lambda_P}}{t}R(t-1)$$

where $\lambda_P = \lambda_{\max}(P^T P)$. Let $\check{A} = B_1 + 2\beta n N_i B_2$. Taking inequalities (4.56), (4.58) and (4.59) into inequality (4.55) gives

$$R(t) \leq \left(1 - \frac{2\beta f_M - \frac{\lambda_n \lambda_Q}{\alpha} - 2\sqrt{\lambda_Q \lambda_P}}{t}\right) R(t-1) + \frac{\alpha}{t} V(t-1) + \frac{\check{A}}{t^2}$$

Finally, it follows that by Lemma 4.3,

$$\lambda_Q = \lambda_P = d_* = \max_{i=1,\cdots,n} d_i$$

which leads to the lemma.

**Lemma 4.6** Under Assumption 4.1, for the state updating Eq. (4.49), there exist positive constants $K$ and $\tilde{A}$ such that the Lyapunov function $V(t)$ follows

$$V(t) \leq \left(1 - \frac{\lambda_2^2}{\lambda_n t}\right) V(t-1) + \frac{2\lambda_n^2 d_*}{\lambda_2^2 t} R(t-1) + \frac{\tilde{A}}{t^2}, \quad t > K \quad (4.60)$$

where $R(t-1)$ is the mean square error of state estimation at time $t-1$, $\lambda_n$ and $\lambda_2$ are respectively the biggest and the smallest positive eigenvalues of Laplacian Matrix $L$, $d_*$ is the maximum degree of the nodes in the system network.

**Proof** By the state updating Eq. (4.49), the Lyapunov function satisfies

$$V(t) = E(x(t)^T L x(t))$$
$$= E\left[x(t-1)^T \left(I - \frac{L}{t}\right) L \left(I - \frac{L}{t}\right) x(t-1)\right] +$$
$$\frac{2}{t} E\left[x(t-1)^T \left(I - \frac{L}{t}\right) L P \varepsilon(t-1)\right] +$$
$$\frac{1}{t^2} E[\varepsilon(t-1)^T P^T L P \varepsilon(t-1)]$$

Since $L = \tilde{L}^T \tilde{L}$, it follows that

$$E\left[x(t-1)^T \left(I - \frac{L}{t}\right) L \varepsilon(t-1)\right]$$
$$= E\left[x(t-1)^T \left(I - \frac{L}{t}\right) \tilde{L}^T \tilde{L} P \varepsilon(t-1)\right]$$
$$\leq \sqrt{E\left[x(t-1)^T \left(I - \frac{L}{t}\right) L \left(I - \frac{L}{t}\right) x(t-1)\right]} \cdot$$
$$\sqrt{E[\varepsilon(t-1)^T P^T L P \varepsilon(t-1)]}$$

By Lemma 4.6, we have

$$E\left[x(t-1)^T\left(I-\frac{L}{t}\right)L\left(I-\frac{L}{t}\right)x(t-1)\right]$$

$$\leq \left(1-\frac{2c_1}{t}+\frac{c_2}{t^2}\right)E[x(t-1)^T Lx(t-1)]$$

$$\leq \left(1-\frac{3c_1}{2t}\right)V(t-1), \text{ if } t > \frac{2c_2}{c_1}$$

Noticing that

$$E(\varepsilon(t-1)^T P^T LP\varepsilon(t-1)) \leq \lambda_n d_* R(t-1)$$

we can obtain, if $t > \max\left\{\frac{2c_2}{c_1}, \frac{2}{3c_1}\right\}$

$$V(t) \leq \left(1-\frac{3c_1}{2t}\right)V(t-1) + \frac{\lambda_n d_*}{t^2}R(t-1) + \frac{2}{t}\sqrt{\lambda_n d_* V(t-1) R(t-1)}$$

$$= \left(1-\frac{3c_1}{2t}\right)V(t-1) + \frac{\lambda_n d_*}{t^2}R(t-1) + \frac{2}{t}\sqrt{\frac{c_1}{2}V(t-1)\frac{2\lambda_n d_*}{c_1}R(t-1)}$$

$$= \left(1-\frac{c_1}{t}\right)V(t-1) + \frac{2\lambda_n d_*}{c_1 t}R(t-1) + \frac{\lambda_n d_*}{t^2}R(t-1)$$

By Lemma 4.4, $c_1 = \frac{\lambda_2^2}{\lambda_n}$. Due to Proposition 4.1 and Remark 4.7, $R(t)$ is bounded as $t > d_*$. Therefore, there exist $K = \max\left\{d_*, \frac{2c_2}{c_1}, \frac{2}{3c_1}\right\}$ and $0 < \tilde{A} < \infty$, such that

$$V(t) \leq \left(1-\frac{\lambda_2^2}{\lambda_n t}\right)V(t-1) + \frac{2\lambda_2^2 d_*}{\lambda_2^2 t}R(t-1) + \frac{\tilde{A}}{t^2}, \quad t > K$$

The lemma is proved.

From Lemma 4.5 and Lemma 4.6, we can see the mean square error $R(t)$ and the Lyapunov function $V(t)$ have a similar form, and they interact with each other. So, we analyze the mean square error and the Lyapunov function together, then we can get the following theorems.

**Theorem 4.6**  Under Assumptions 4.1 and 4.2, we have the following results for Consensus Algorithm,

## 4.3 Recursive Projection Consensus

$$E(\hat{x}_{ij}(t) - x_j(t))^2 = \begin{cases} O(\dfrac{1}{t^{\lambda_{\min}(W)}}) & \lambda_{\min}(W) < 1 \\ O\left(\dfrac{\ln t}{t}\right) & \lambda_{\min}(W) = 1 \\ O\left(\dfrac{1}{t}\right) & \lambda_{\min}(W) > 1 \end{cases} \quad (4.61)$$

for $j \in N_i$, $i = 1, \cdots, n$ and

$$E(x_i(t) - x_j(t))^2 = \begin{cases} O(\dfrac{1}{t^{\lambda_{\min}(W)}}) & \lambda_{\min}(W) < 1 \\ O\left(\dfrac{\ln t}{t}\right) & \lambda_{\min}(W) = 1 \\ O\left(\dfrac{1}{t}\right) & \lambda_{\min}(W) > 1 \end{cases} \quad (4.62)$$

for $\forall i, j = 1, \cdots, n, i \neq j$, where

$$W = \begin{pmatrix} \dfrac{\lambda_2^2}{\lambda_n} & -\dfrac{2\lambda_n^2 d_*}{\lambda_2^2} \\ -\dfrac{2\lambda_n^2 d_*}{\lambda_2^2} & 2\beta f_M - \dfrac{\lambda_2^2}{2\lambda_n} - 2d_* \end{pmatrix}$$

$f_M = f(|C| + M)$, $\lambda_n$ and $\lambda_2$ are respectively the biggest and the smallest positive eigenvalues of Laplacian Matrix $L$, $d_*$ is the maximum degree of the nodes in the system network.

**Proof** Considering the mean square error Eq. (4.52) and the Lyapunov function Eq. (4.60) together, we have

$$\begin{cases} V(t) \leq \left(1 - \dfrac{\lambda_2^2}{\lambda_n t}\right) V(t-1) + \dfrac{2\lambda_n^2 d_*}{\lambda_2^2 t} R(t-1) + \dfrac{\tilde{A}}{t^2} \\ R(t) \leq \left(1 - \dfrac{2\beta f_M - \dfrac{\lambda_n d_*}{\alpha} - 2d_*}{t}\right) R(t-1) + \\ \dfrac{\alpha}{t} V(t-1) + \dfrac{\check{A}}{t^2} \end{cases} \quad (4.63)$$

Let $\alpha = \dfrac{\lambda_2^2}{\lambda_n}$, $b = -\dfrac{2\lambda_n^2 d_*}{\lambda_2^2}$, $c = 2\beta f_M - \dfrac{\lambda_n d_*}{\alpha} - 2d_*$, $d = -\alpha$.

Denote

$$Z(t) = \begin{pmatrix} V(t) \\ R(t) \end{pmatrix}, \quad W = \begin{pmatrix} a & b \\ d & c \end{pmatrix}, \quad H = \begin{pmatrix} \tilde{A} \\ \check{A} \end{pmatrix}$$

By Eq. (4.63), we have

$$\|Z(t)\| \leq \left\|\left(I - \frac{W}{t}\right)Z(t-1) + \frac{H}{t^2}\right\| \leq \left\|I - \frac{W}{t}\right\| \|Z(t-1)\| + \frac{\|H\|}{t^2} \tag{4.64}$$

Let $\alpha = \dfrac{2\lambda_n^2 d_*}{\lambda_2^2}$, then $b = d$. The matrix $W$ is a symmetric matrix, and then,

$$\lambda\left(I - \frac{2W}{t} + \frac{W^2}{t^2}\right) = \left(1 - \frac{\lambda(W)}{t}\right)^2 \leq \left(1 - \frac{\lambda_{\min}(W)}{t}\right)^2, \quad \text{as} \quad t > \lambda_{\max}(W)$$

It follows that

$$\left\|I - \frac{W}{t}\right\| = \sqrt{\lambda_{\max}\left(I - \frac{2W}{t} + \frac{W^2}{t^2}\right)} \leq 1 - \frac{\lambda_{\min}(W)}{t}$$

Consequently, we have by Eq. (4.64),

$$\|Z(t)\| \leq \left(1 - \frac{\lambda_{\min}(W)}{t}\right)\|Z(t-1)\| + \frac{\|H\|}{t^2}$$

$$\leq \prod_{i=1}^{t}\left(1 - \frac{\lambda_{\min}(W)}{i}\right)\|Z(0)\| + \sum_{i=1}^{t}\prod_{l=i+1}^{t}\left(1 - \frac{\lambda_{\min}(W)}{t}\right)\frac{\|H\|}{i^2}$$

According to Theorem 2.10, it follows that

$$\|Z(t)\| = \begin{cases} O\left(\dfrac{1}{t^{\lambda_{\min}(W)}}\right) & \lambda_{\min}(W) < 1 \\ O\left(\dfrac{\ln t}{t}\right) & \lambda_{\min}(W) = 1, \quad \text{as } t \to \infty \\ O\left(\dfrac{1}{t}\right) & \lambda_{\min}(W) > 1 \end{cases} \tag{4.65}$$

Since the network $G$ is connected, there exists a road between any different agents $i$ and $j$. Suppose the road is as follows

$$i = r_0 \to r_1 \to r_2 \to \cdots \to r_{p-1} \to r_p = j, \quad p \leq n$$

which implies $r_{i+1} \in N_{r_i}$. Then the mean square error of any two different agents follows

$$E(x_i(t) - x_j(t))^2$$

$$= E[(x_{r_0}(t) - x_{r_1}(t)) + (x_{r_1}(t) - x_{r_2}(t)) + \cdots + (x_{r_{p-1}}(t) - x_{r_p}(t))]^2$$

$$\leq p \sum_{i=1}^{p-1} E(x_{r_i}(t) - x_{r_{i+1}}(t))^2$$

$$\leq n \sum_{i=1}^{n} \sum_{j \in N_i} E(x_i(t) - x_j(t))^2$$

It is easy to verify

$$R(t) = E[e(t)^T e(t)] = \sum_{i=1}^{n} \sum_{j \in N_i} E(\hat{x}_{ij}(t) - x_j(t))^2$$

and

$$V(t) = E[x^T(t) L x(t)] = \frac{1}{2} \sum_{i=1}^{n} \sum_{j \in N_i} E(x_i(t) - x_j(t))^2$$

Thus,

$$E(\hat{x}_{ij}(t) - x_j(t))^2 \leq R(t), \quad E(x_i(t) - x_j(t))^2 \leq 2nV(t)$$

What's more,

$$R(t) \leq |Z(t)|, \quad V(t) \leq |Z(t)|$$

then the theorems follows by Eq. (4.65).

**Theorem 4.7** (Weak Consensus) Let $\gamma_1 = \dfrac{4\lambda_n^5 d_*^2}{\lambda_2^6} + \dfrac{\lambda_2^2}{2\lambda_n} + 2d_* $ and $\gamma_2 = \dfrac{4\lambda_n^5 d_*^2}{\lambda_2^4 (\lambda_2^2 - \lambda_n)} + \dfrac{\lambda_2^2}{2\lambda_n} + 2d_* + 1$. Under Assumptions 4.1 and 4.2, the following results about Consensus Algorithm hold:

(1) If $\beta > \dfrac{\gamma_1}{2 f_M}$, then the estimates of the states will converge to the true states as $t \to \infty$:

$$E(\hat{x}_{ij}(t) - x_j(t))^2 \to 0, \quad j \in N_i, \quad i = 1, \cdots, n \tag{4.66}$$

and the states will achieve weak consensus:

$$E(x_i(t) - x_j(t))^2 \to 0, \quad \forall i, j = 1, \cdots, n, \quad i \neq j \tag{4.67}$$

(2) If $\dfrac{\lambda_2^2}{\lambda_n} > 1$ and $\beta > \dfrac{\gamma_2}{2 f_M}$, then as $t \to \infty$

$$E(\hat{x}_{ij}(t) - x_j(t))^2 = O\left(\frac{1}{t}\right), \quad j \in N_i, \quad i = 1, \cdots, n$$

and

$$E(x_i(t) - x_j(t))^2 = O\left(\frac{1}{t}\right), \quad j \in N_i, \quad i = 1, \cdots, n$$

The notations $\beta, f_M, \lambda_2, \lambda_n, d_*$ are all the same as that in Theorem 4.6.

**Proof** The matrix $W$ defined in Theorem 4.6 is

$$W = \begin{pmatrix} a & b \\ d & c \end{pmatrix}$$

where $a = \dfrac{\lambda_2^2}{\lambda_n}$, $b = d = -\dfrac{2\lambda_n^2 d_*}{\lambda_2^2}$, $c = 2\beta f_M - \dfrac{\lambda_2^2}{2\lambda_n} - 2d_*$.

Letting $|\lambda I - W| = (\lambda - a)(\lambda - c) - b^2 = 0$, we can get the eigenvalues of matrix $W$ are

$$\lambda_{max}(W) = \frac{a + c + \sqrt{(a+c)^2 - 4(ac - b^2)}}{2}$$

$$\lambda_{min}(W) = \frac{a + c - \sqrt{(a+c)^2 - 4(ac - b^2)}}{2} \qquad (4.68)$$

(1) If $\beta > \dfrac{\gamma_1}{2f_M}$, then

$$c = 2\beta f_M - \frac{\lambda_2^2}{2\lambda_n} - 2d_* > \frac{4\lambda_n^5 d_*^2}{\lambda_2^6} > 0$$

$$ac > \frac{\lambda_2^2}{\lambda_n} \frac{4\lambda_n^5 d_*^2}{\lambda_2^6} = b^2$$

The smallest eigenvalue of matrix $W$ follows

$$\lambda_{min}(W) = \frac{a + c - \sqrt{(a+c)^2 - 4(ac - b^2)}}{2} > 0$$

By Eq. (4.61) and Eq. (4.62), we can get

$$E(\hat{x}_{ij}(t) - x_j(t))^2 \to 0, \quad j \in N_i, \quad i = 1, \cdots, n$$

and

$$E(x_i(t) - x_j(t))^2 \to 0, \quad \forall i, j = 1, \cdots, n, \quad i \neq j$$

(2) If $\beta > \dfrac{\gamma_2}{2f_M}$, then

$$c = 2\beta f_M - \frac{\lambda_2^2}{2\lambda_n} - 2d_* > \frac{4\lambda_n^5 d_*^2}{\lambda_2^4(\lambda_2^2 - \lambda_n)} + 1 = \frac{b^2}{a-1} + 1$$

If $a = \dfrac{\lambda_2^2}{\lambda_n} > 1$, then

$$c(a-1) > b^2 + (a-1)$$

Consequently,

$$ac - b^2 > a + c - 1$$

and

$$(a+c)^2 - 4(ac - b^2) < (a+c)^2 - 4(a+c) + 4 = (a+c-2)^2$$

which implies

$$\sqrt{(a+c)^2 - 4(ac - b^2)} < a + c - 2$$

Hence, the smallest eigenvalue of matrix $W$ follows

$$\lambda_{\min}(W) = \frac{(a+c) - \sqrt{(a+c)^2 - 4(ac - b^2)}}{2} > \frac{(a+c) - (a+c-2)}{2} = 1$$

By Eq. (4.61) and Eq. (4.62) again, we can get the second result of the theorem.

**Theorem 4.8** (Mean Square Consensus) Under Assumptions 4.1 and 4.2, there exists a random variable

$$x^* = \frac{1}{n} \mathbb{1}^T x(1) + \frac{1}{n} \sum_{i=1}^{\infty} \frac{1}{i} \mathbb{1}^T P \varepsilon(i)$$

where $\mathbb{1}$ is the $n$-dimensional vector with all the elements being 1, such that the following results about Consensus Algorithm hold:

(1) If $\beta > \dfrac{\gamma_1}{2f_M}$, then the states in the system can achieve mean square consensus:

$$E(x_i(t) - x^*)^2 \to 0, \quad \forall i = 1, \cdots, n, \text{ as } t \to \infty$$

(2) If $\dfrac{\lambda_2^2}{\lambda_n} > 1$ and $\beta > \dfrac{\gamma_2}{2f_M}$, then the convergence rate of mean square consensus is

$$E(x_i(t) - x^*)^2 = O\left(\frac{1}{t}\right), \quad \forall i = 1, \cdots, n, \text{ as } t \to \infty$$

The notations $\beta, f_M, \lambda_2, \lambda_n, \gamma_1, \gamma_2$ are all the same as that in Theorem 4.7.

**Proof** (1) Since $\mathbb{1}^T L = 0$, we have by the state updating (4.49)

$$\mathbb{1}^T x(t+1) = \mathbb{1}^T x(t) + \frac{1}{t} \mathbb{1}^T P \varepsilon(t) = \mathbb{1}^T x(1) + \sum_{i=1}^{t} \frac{1}{i} \mathbb{1}^T P \varepsilon(i)$$

For any $m \in Z^+$, we have

$$E(\mathbb{1}^T x(t+m) - \mathbb{1}^T x(t))^2 = E\left(\sum_{i=t}^{t+m-1} \frac{1}{i} \mathbb{1}^T P \varepsilon(i)\right)^2$$

$$= E\left(\sum_{i=t}^{t+m-1}\sum_{j=t}^{t+m-1}\frac{1}{i}\frac{1}{j}\mathbb{I}^T P\varepsilon(i)\,\mathbb{I}^T P\varepsilon(j)\right)$$

$$\leq \sum_{i=t}^{t+m-1}\sum_{j=t}^{t+m-1}\frac{1}{i}\frac{1}{j}\sqrt{E(\mathbb{I}^T P\varepsilon(i))^2 E(\mathbb{I}^T P\varepsilon(j))^2}$$

If $\beta > \dfrac{\gamma_1}{2f_M}$, then $\lambda_{\min}(W) > 0$. Consequently, we have by Eq. (4.61)

$$E(\mathbb{I}^T x(t+m) - \mathbb{I}^T x(t))^2 = O\left(\sum_{i=t}^{t+m-1}\frac{1}{i}\sqrt{R(i)}\right)^2 \to 0$$

According to Theorem 2.9, there exists a random variable $S_\infty \triangleq \mathbb{I}^T x(1) + \sum_{i=1}^{\infty}\dfrac{1}{i}\mathbb{I}^T Pe(i)$ such that

$$E(\mathbb{I}^T x(t) - S_\infty)^2 \to 0, \text{ as } t \to \infty \tag{4.69}$$

which means $\mathbb{I}^T x(t)$ converges to $S_\infty$ in the mean square sense.

Let $x^* = \dfrac{S_\infty}{n}$, then

$$E(x_i(t) - x^*)^2 = E\left(x_i(t) - \frac{S_\infty}{n}\right)^2$$

$$= \frac{1}{n^2}E(nx_i(t) - \mathbb{I}^T x(t) + \mathbb{I}^T x(t) - S_\infty)^2$$

$$= \frac{1}{n^2}E\left(\sum_{j=1, j\neq i}^{n}(x_i(t) - x_j(t)) + (\mathbb{I}^T x(t) - S_\infty)\right)^2$$

$$\leq \frac{1}{n}\sum_{j=1, j\neq i}^{n}E(x_i(t) - x_j(t))^2 + \frac{2}{n}E(\mathbb{I}^T x(t) - S_\infty)^2 \tag{4.70}$$

Since $E(x_i(t) - x_j(t))^2 \to 0$ by Eq. (4.67) and $\mathbb{I}^T x(t)$ converges to $S_\infty$ by Eq. (4.69) in mean square sense, it follows that $x_i(t)$ converges to $x^*$ in mean square sense.

(2) If $\dfrac{\lambda_2^2}{\lambda_n} > 1$ and $\beta > \dfrac{\gamma_2}{2f_M}$, we have by Theorem 4.7

$$E(x_i(t) - x_j(t))^2 = O\left(\frac{1}{t}\right), \ j \in N_i, \ i = 1, \cdots, n \tag{4.71}$$

and

$$R(t) = E\varepsilon(t)^T \varepsilon(t) = O\left(\frac{1}{t}\right) \tag{4.72}$$

## 4.3 Recursive Projection Consensus

According to inequality (4.70), the second result in the theorem will be proved if we can show

$$E(\mathbb{I}^T x(t) - S_\infty)^2 = O\left(\frac{1}{t}\right) \tag{4.73}$$

Since

$$E(\mathbb{I}^T x(t) - S_\infty)^2 = E\left(\sum_{i=t+1}^\infty \frac{1}{i} \mathbb{I}^T P\varepsilon(i)\right)^2 \leq E\left(\sum_{i=t+1}^\infty \sum_{j=t+1}^\infty \frac{1}{ij} |\varepsilon(i)^T P^T \mathbb{I}\mathbb{I}^T P\varepsilon(j)|\right)$$

We have by [[30], Corollary 4.1.2]

$$E(\mathbb{I}^T x(t) - S_\infty)^2$$

$$\leq \sum_{i=t+1}^\infty \sum_{j=t+1}^\infty \frac{1}{ij} E|\varepsilon(i)^T P^T \mathbb{I}\mathbb{I}^T P\varepsilon(j)|$$

$$\leq \sum_{i=t+1}^\infty \sum_{j=t+1}^\infty \frac{1}{ij} \sqrt{E[\varepsilon(i)^T P^T \mathbb{I}\mathbb{I}^T P\varepsilon(i)]} \sqrt{E[\varepsilon(j)^T P^T \mathbb{I}\mathbb{I}^T P\varepsilon(j)]}$$

$$= O\left(\sum_{i=t+1}^\infty \frac{1}{i} \sqrt{R(i)}\right)^2$$

Together with Eq. (4.72), we have

$$E(\mathbb{I}^T x(t) - S_\infty)^2 = O\left(\frac{1}{i^{1+1/2}}\right) = O\left(\frac{1}{t}\right)$$

which leads to the theorem.

**Remark 4.8** The conditions in Theorem 4.7 and Theorem 4.8 are sufficient but not necessary.

**Remark 4.9** For any $x_i \in R$, $i = 1, \cdots, n$, weak consensus follows by mean square consensus, but mean square consensus cannot be derived from weak consensus. Since the states are updated by Eq. (4.49) and the estimates have good quality in our algorithm, the states can achieve mean square consensus by weak consensus.

**Remark 4.10** Compared with the two-time-scale algorithm in Section 4.2, the convergence rate in this section is more faster. In two-time-scale, the convergence rate is $O(1/t)$ in small time scale with the best case, but in large time scale the convergence rate cannot achieve $O(1/t)$ due to long time keeping the states unchanged. However, the convergence rate of the algorithm in this chapter can achieve $O(1/t)$.

### 4.3.3 Numerical Simulation

In the simulations, we consider the network with three agents as Fig. 4.6. We can see the sets of neighbors are as follows

$$N_1 = \{2, 3\}, \quad N_2 = \{1, 3\}, \quad N_3 = \{1, 2\}$$

where $N_i$ is the set of agent $i$'s neighbors. The corresponding matrixes $P$, $Q$ and Laplacian Matrix $L$ are given in Eq. (4.50) and Eq. (4.51). The biggest and the smallest positive eigenvalues of Laplacian Matrix $L$ are respectively $\lambda_3 = 3$ and $\lambda_2 = 3$. Then, the parameter $\gamma_1$ in conditions of Theorem 4.7 and Theorem 4.8 can be calculated by $\gamma_1 = \dfrac{4\lambda_3^5 d_*^2}{\lambda_2^6} + \dfrac{\lambda_2^2}{2\lambda_3} + 2d_* = 10.8333$.

The states of the agents in the network are updated by

$$x_i(t+1) = x_i(t) + u_i(t), \quad i = 1, 2, 3 \tag{4.74}$$

The control $u_i(t)$ is designed by

$$u_i(t) = -\frac{1}{t+1}\sum_{j \in N_i}(x_i(t) - \hat{x}_{ij}(t)), \quad i = 1, 2, 3$$

where $\hat{x}_{ij}(t)$ is the estimate of agent $j$'s state by agent $i$ at time $t$. It can be calculated by

$$\hat{x}_{ij}(t) = \prod_M \left\{ \hat{x}_{ij}(t-1) + \frac{\beta}{t}(F(C - \hat{x}_{ij}(t-1)) - s_{ij}(t)) \right\} \tag{4.75}$$

where $j \in N_i$, $i = 1, 2, 3$, $s_{ij}(t)$ is the binary-valued measurement and $\prod_M(\cdot)$ is the projection operator.

Let the initial states, the estimates, the threshold, and the state bound are respectively

$$x(0) = [x_1(0), x_2(0), x_3(0)]^T = [-5, 1, 8]^T,$$

$$\hat{x}(0) = [\hat{x}_{12}(0), \hat{x}_{13}(0), \hat{x}_{21}(0), \hat{x}_{22}(0), \hat{x}_{31}(0), \hat{x}_{32}(0)]^T$$
$$= [0, 0, 2, 2, -3, -3]^T$$

$C = 0$ and $M = 8$. The distribution of the noise is $N(0, 6)$. Then $f_M = f(|C| + M) = 0.0273$. Let $\beta = 1$ in Eq. (4.75), the estimates $\hat{x}_{ij}(t)$, $j \in N_i$, $i = 1, 2, 3$ are updated as Fig. 4.7(a). Based on the estimates, the states $x_i(t)$, $i = 1, 2, 3$ are updated by Eq. (4.74), which are shown in Fig. 4.7(b). In Fig. 4.7, $\beta = 1 < \dfrac{\gamma_1}{2f_M} = 198.41$, which does not satisfy the condition in Theorem 4.6, and it is shown that neither the states achieve consensus nor the estimates converge to the true states.

Fig. 4.8 and Fig. 4.9 are respectively the cases with $\beta = 20$ and $\beta = 200$. In Fig. 4.8, $\beta = 20 < \dfrac{\gamma_1}{2f_M}$, which does not satisfy the condition in Theorem 4.7 (1) and Theorem 4.8 (1), but it is shown that the estimates can converge to the true states, and the sates can achieve weak consensus and mean square consensus. This illustrate that the

4.3 Recursive Projection Consensus

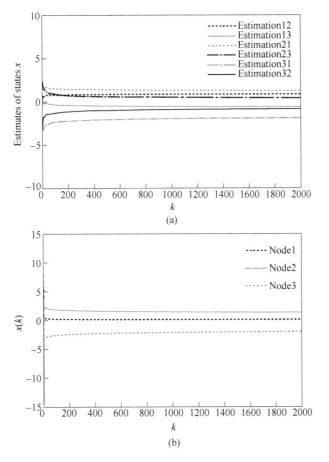

Fig. 4.7 $\beta = 1$

(a) The estimates; (b) State updating

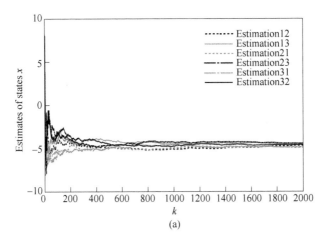

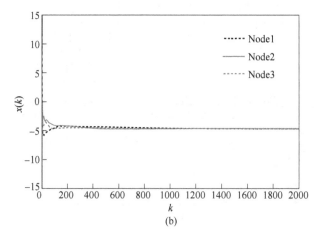

Fig. 4.8  $\beta = 20$

(a) The estimates; (b) State updating

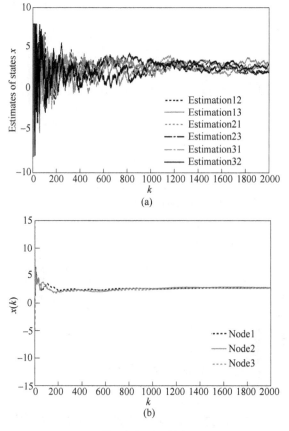

Fig. 4.9  $\beta = 200$

(a) The estimates; (b) State updating

condition in Theorem 4.7 (1) and Theorem 4.8 (1) is not necessary. In Fig. 4.8, $\beta = 200 > \frac{\gamma_1}{2f_M}$, the states can achieve weak consensus and mean square consensus, too. The results are in accord with Theorem 4.7 and Theorem 4.8.

Fig. 4.10 is a comparison of the results of two-time-scale algorithm in Section 4.2. We choose the holding time $L_k = k^\alpha = k^2$ in the two-time-scale algorithm and $\beta = 20$ in the algorithm in the recursive projection algorithm. Fig. 4.10 is the comparison of the variance of the states. We can see that the convergence rate of the recursive projection algorithm is faster than that of two-time-scale algorithm, which is consistent with Remark 4.10.

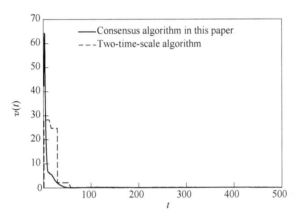

Fig. 4.10  Comparison with the results of two-time-scale algorithm

## 4.4  Notes

This chapter studies the consensus control of multi-agent systems under binary-valued measurements. Based on the non-truncated identification method and the recursive projection identification method, two time-scale consensus algorithms and recursive projection consensus algorithm are designed, respectively. The differences between the two algorithms are as follows. There is a waiting time in the two-time-scale consensus algorithm, during which the binary information of the agents' states is collected for estimation and then control is designed. There is no waiting time in the recursive projection consensus algorithm. The states are estimated online and the control is also carried out at every moment. For these two consensus algorithms, the convergence and convergence rate are given, respectively.

# 5 Consensus with Binary-Valued Measurements under Directed Topology

For some consensus problems, the states are required to converge to the average of the initial states of the entire team, which induces an average consensus problem. The average consensus of the multi-agent system has been a hot research topic.

In this chapter, we study the average consensus problem with binary-valued measurements under directed topologies. In the directed topologies, a node can only get the binary information from its parent nodes, and cannot get the information from its child nodes. As a result, the information that each node gets is less, comparing with undirected topologies. Besides, average consensus is more difficult to achieve than the general consensus. Thus, the average consensus of multi - agent systems under directed topologies is to complete a more difficult task by using less information, which is more challenging.

The structure of this chapter is as follows: Section 5.1 decribes the problem formulation. Section 5.2 introduces a stochastic approximation based consensus algorithm. Section 5.3 gives the properties of the algorithm, including the convergence and convergence rate. Simulations are given to demonstrate the theoretical results in Section 5.4. The concluding remarks and some related future topics are given in Section 5.5.

## 5.1 Problem Formulation

Consider a multi-agent system with $n$ nodes as follows.

$$x_i(t+1) = x_i(t) + u_i(t), \quad i = 1, \cdots, n \tag{5.1}$$

or in a vector form:

$$x(t+1) = x(t) + u(t) \tag{5.2}$$

where $x(t) = [x_1(t), \cdots, x_n(t)]^T$ and $u(t) = [u_1(t), \cdots, u_n(t)]^T$, $x_i(t) \in R$ is agent $i$'s state at time $t$, $u_i(t) \in R$ is the control of agent $i$ at time $t$.

The agents in system (5.1) interact with each other through an information network, which is given by a directed graph $G_I = \{N, E\}$ consisting of $n$ agents and $l_s$ links. Denote each edge as an ordered pair $(i, j)$ where $i \neq j$. If an edge $(i, j) \in G_I$, then agent $j$ can obtain the information from agent $i$ and agent $i$ is called a neighbor of agent $j$.

For the edge $(i, j) \in G_I$, $i$ is called the parent agent and $j$ is called the child agent.

The information each agent $i$ gets from its neighbors $\{j, (j, i) \in G_I\}$ is measured by a binary-valued sensor with a threshold $C$ as follows:

$$\begin{cases} y_{ij}(t) = x_j(t) + d_{ij}(t) \\ s_{ij}(t) = I_{|y_{ij}(t) \leq C|} = \begin{cases} 1 & \text{if } y_{ij}(t) \leq C \\ 0 & \text{if } y_{ij}(t) > C \end{cases} \end{cases} \quad (5.3)$$

where $x_j(t)$ is the state of agent $j$, $d_{ij}(t)$ is system noise, $y_{ij}(t)$ is the output data which cannot be measured, and $s_{ij}(t)$ is the binary-valued observations, $C$ is a given threshold of the binary-valued sensor.

The goal is to design the control $u_i(t)$ based on the binary information $s_{ij}(t)$ to achieve average consensus. To realize $u_i(t)$, each agent $i$ designs the quantity of transportation $v_{ij}(t)$, which results in a loss of $v_{ij}(t)$ at agent $i$ and a gain of $v_{ij}(t)$ at agent $j$. Then $u_i(t)$ will be realized by

$$u_i(t) = -\sum_{j \neq i, j=1}^{n} v_{ij}(t) + \sum_{j \neq i, j=1}^{n} v_{ji}(t) \quad (5.4)$$

In fact, what agent $i$ designs is the quantity of transportation $v_{ij}(t)$, $j = 1, \cdots, n$, rather than $u_i(t)$, which means $v_{ij}(t)$ is the real control signal.

Denote $\bar{x} = \frac{1}{n} \sum_{i=1}^{n} x_i(0)$. For the multi-agent system (5.1) and (5.4), the transportation $v_{ij}(t)$ is designed by agent $i$ based on its state $x_i(t)$ and its observations $s_{ij}(t)$ to achieve convergence of $x(t)$ towards $\bar{x} \mathbb{1}_n$, where $\mathbb{1}_n$ is the $n$ dimensional column vector with all elements being 1.

To solve the problem, we have the following assumptions:

**Assumption 5.1** The information network $G_I$ is strong connected.

**Assumption 5.2** The noises $\{d_{ij}(t), (j, i) \in G_I, t = 1, 2, \cdots\}$ are independent with respect to $i, j, t$ and identically distributed. The distribution function $F(\cdot)$ is invertible and twice continuously differentiable, and the associated density function satisfies $f(x) = dF(x)/dx \neq 0$.

**Assumption 5.3** The quantity of transportation $v_{ij}$ is designed by agent $i$ and its value is available to agent $j$. All others can neither influence $v_{ij}(t)$ nor know its value.

## 5.2 Control Algorithm

As we all know, the consensus law is designed by using the true states of the neighbors if the neighbors' states can be obtained. However, the states of the neighbors' cannot be obtained in this book. A straightforward idea is to replace the real state with an estimate. Each agent should estimate its neighbors' states firstly and then design the control based on the estimation. So, we use the idea of two-time-scale algorithm in Section 4.2 to design the average consensus control for the multi-agent systems with directed topology.

For agent $i$, it begins to estimate its neighbors $j$'s states at the time $t_k$. It takes $L_k$ time to accumulate the binary-valued information from its neighbors. Based on the binary-valued information, agent $i$ estimates its neighbors' states. Based on the estimates, agent $i$ designs the quantity of transportation $v_{ij}(t_{k+1}-1)$ at time $t_{k+1}-1=t_k+L_k-1$, which makes the control $u_i(t_{k+1}-1)$ be realized by Eq. (5.4) and the states be updated by Eq. (5.1) at time $t_{k+1}-1$. Based on the new states, the process alternating estimation and control will be repeated.

In general, the more information an agent gets from its neighbors, the more accurately the agent estimates its neighbors' states. Therefore, the holding time for estimation is designed to be an increasing function $L_k$. In this chapter, we choose $L_k = k^\beta$, $\beta > 0$. Assume there exists a constant $B$ such that $\sum_{i=1}^{n} |x_i(0)| \leq B$. The Average Consensus Algorithm (ACA) is constructed as follows:

(1) To set initial conditions: Let $k=1$ and $t_1 = 0$.

(2) To get the binary-valued measurements:

Let $t_{k+1} = t_k + L_k$, where $L_k = k^\beta$, $\beta > 0$. In the time interval $[t_k, t_{k+1})$, all the states keep unchanged,

$$x_i(t_k) = x_i(t_k+1) = \cdots = x_i(t_{k+1}-1) \triangleq z_i(k)$$

And, each agent $i$ accumulates $L_k$ binary-valued measurements from its neighbors:

$$\begin{cases} y_{ij}(l) = x_j(t_k) + d_{ij}(l) \\ s_{ij}(l) = I_{\{y_{ij}(l) \leq C\}} \end{cases}, \sim (j, i) \in G_I$$

for $l = t_k, \cdots, t_{k+1}-1$.

(3) To estimate state: For agent $i$, the frequency that binary-valued measurements are taken from its neighbors for the holding time $L_k$ can be given by

$$\varphi_{ij}(L_k) = \frac{1}{L_k} \sum_{l=t_k}^{t_{k+1}-1} s_{ij}(l)$$

Modifying the frequency $\varphi_{ij}(L_k)$ by a truncation $a$, we have $\xi_{ij}(L_k)$ as follows:

$$\xi_{ij}(L_k) = \begin{cases} \varphi_{ij}(L_k) & \text{if } a < \varphi_{ij}(L_k) < 1-a \\ a & \text{if } \varphi_{ij}(L_k) \leq a \\ 1-a & \text{if } \varphi_{ij}(L_k) \geq 1-a \end{cases}$$

## 5.2 Control Algorithm

where

$$a = \min\left\{1 - F\left(C + B + \sum_{h=1}^{\infty} \frac{1}{h^{1+\gamma}}\right), F\left(C - B - \sum_{h=1}^{\infty} \frac{1}{h^{1+\gamma}}\right)\right\}, \quad 0 < \gamma < \frac{\beta}{2} \tag{5.5}$$

At last, each agent $i$ estimates its neighbors' states by

$$\hat{x}_{ij}(L_k) = C - F^{-1}(\xi_{ij}(L_k)), \quad (j, i) \in G_I \tag{5.6}$$

(4) To design the quantity of the transportation:

1) To design the quantity of the first transportation:

Based on the estimation $\hat{x}_{ij}(L_k)$, agent $i$ designs the quantity of the first transportation $v_{ij}^1(k)$ by

$$v_{ij}^1(k) = \begin{cases} \frac{1}{k}(x_i(t_k) - \hat{x}_{ij}(L_k)) & (j, i) \in G_I \\ 0 & \text{others} \end{cases}$$

Then, the first state updating by $v_{ij}^1(k)$ is as follows

$$x_i^1(t_{k+1}) = x_i(t_k) - \sum_{(j, i) \in G_I} v_{ij}^1(k) + \sum_{(i, j) \in G_I} v_{(j, i)}^1(k) \tag{5.7}$$

2) To design the quantity of the second transportation:

Each agent $i$ checks whether the first state updating satisfies the following inequality:

$$|x_i^1(t_{k+1})| < B + \sum_{h=1}^{k+1} \frac{1}{h^{1+\gamma}} \tag{5.8}$$

If there exists an agent $i_1$ who doesn't meet the condition Eq. (5.8), then it passes an alarm information to its child agents $\{i_2, (i_1, i_2) \in G_I\}$ that there is a mistake in the first transportation (assume it takes 1 unit time to pass the information), and designs the second transportation as follows

$$v_{i_1 j}^2(k) = - v_{i_1 j}^1(k), \quad j \neq i_1 \tag{5.9}$$

Then, the agents $i_3$ who receive the alarm information for the first time pass the information to their child agents $\{i_4, (i_3, i_4) \in G_I\}$, and design the second transportation as follows

$$v_{i_3 j}^2(k) = - v_{i_3 j}^1(k), \quad j \neq i_3 \tag{5.10}$$

The process will be repeated until no one is getting the information for the first time.

If an agent doesn't obtain the alarm information for $n$ unit times, it designs the second transportation as

$$v_{ij}^2(k) = 0 \tag{5.11}$$

3) To design the quantity of transportation: The quantity of transportation at time $t_{k+1}-1$ is designed by

$$v_{ij}(t_{k+1} - 1) = v_{ij}^1(k) + v_{ij}^2(k)$$

The states of the agents will be updated by

$$x_i(t_{k+1}) = x_i(t_k) - \sum_{j \neq i,\, j=1}^{n} v_{ij}(t_{k+1} - 1) + \sum_{j \neq i,\, j=1}^{n} v_{ji}(t_{k+1} - 1)$$

(5) $k = k + 1$, go back to step (2).

**Remark 5.1** Under Assumption 5.1, the alarm information will spread across the entire network if there is an agent that does not satisfy condition Eq. (5.8), which will take up to $n$ unit times. All the agents ($i=1, \cdots, n$) will design the second transportation as Eq. (5.9) and Eq. (5.10), that is to say $v_{ij}^2(k) = - v_{ij}^1(k)$. Thus, all the quantity of transportation will be zero and the states of the agents will be updated by

$$x_i(t_{k+1}) = x_i(t_k) \tag{5.12}$$

If an agent doesn't obtain the alarm information for $n$ unit times, which implies all the agents meet condition Eq. (5.8), then the second transportation will be zero. The first transportation will be the only transportation and the states will be updated by

$$x_i(t_{k+1}) = x_i(t_k) - \frac{1}{k} \sum_{(j,\, i) \in G_I} (x_i(t_k) - \hat{x}_{ij}(L_k)) + \frac{1}{k} \sum_{(i,\, j) \in G_I} (x_j(t_k) - \hat{x}_{ji}(L_k))$$

$$\tag{5.13}$$

**Remark 5.2** Since $|x_i(0)| \leq B$, we can get the following assertions by state updating laws (5.12), (5.13) and mathematical induction:

$$|x_i(t_k)| \leq B + \sum_{h=1}^{\infty} \frac{1}{h^{1+\gamma}} \triangleq B_{\infty},\ k = 1, 2, \cdots$$

for all the agents $i=1, \cdots, n$. As a result, the condition $a < F(C - x_j(t_k)) < 1 - a$ is satisfied.

## 5.3 Properties of the Algorithm

For the sake of simplicity, we denote $z_i(k) = x_i(t_k)$, $i=1, \cdots, n$, and $z(k) = x(t_k) = (x_1(t_k), \cdots, x_n(t_k))^T$. Then, we can get the following results on estimation and transportation design.

### 5.3.1 Estimation

In the Average Consensus Algorithm (ACA) proposed in Section III, the state of each node is constant in the time interval $[t_k, t_{k+1})$. Limiting $t \in [t_k, t_{k+1})$ in Eq. (5.3),

## 5.3 Properties of the Algorithm

we have:

$$\begin{cases} y_{ij}(l) = z_j(k) + d_{ij}(l) \\ s_{ij}(l) = I_{|y_{ij}(l) \leq C|} \end{cases} \quad (j, i) \in G_I, \ l = t_k, \cdots, t_{k+1} - 1 \quad (5.14)$$

where $z_j(k)$ is constant and its estimation $\hat{x}_{ij}(L_k)$ is given by Eq. (5.6).

**Theorem 5.1** Under Assumption 5.2, the estimate Eq. (5.6) in the step (3) of the ACA can converge to the true state:

$$\hat{x}_{ij}(L_k) \to z_j(k), \text{ for any } (j, i) \in G_I, \text{ as } k \to \infty \quad (5.15)$$

**Proof** Under Assumption 5.2, we have by Eq. (5.14)

$$E(s_{ij}(l)) = P(s_{ij}(l) = 1) = P(y_{ij}(l) \leq C) = F(C - z_j(k))$$

Denote $p = F(C - z_j(k))$. Then, by Remark 5.2, one can get

$$a < p < 1 - a$$

According to the strong law of large numbers, we have:

$$\varphi_{ij}(L_k) = \frac{1}{L_k} \sum_{l=t_k}^{t_{k+1}-1} s_{ij}(l) \to p, \text{ a. s. } \text{ as } L_k \to \infty$$

Since $F(\cdot)$ is not invertible at 0 and 1, $\xi_{ij}(L_k)$ is the modification of $\varphi_{ij}(L_k)$:

$$\xi_{ij}(L_k) = \begin{cases} \varphi_{ij}(L_k) & \text{if } a < \varphi_{ij}(L_k) < 1 - a \\ a & \text{if } \varphi_{ij}(L_k) \leq a \\ 1 - a & \text{if } \varphi_{ij}(L_k) \geq 1 - a \end{cases}$$

Noting $\alpha < p < 1 - \alpha$, we have

$$\xi_{ij}(L_k) \to p, \text{ a. s. } \text{ as } k \to \infty$$

Therefore

$$\hat{x}_{ij}(L_k) = C - F^{-1}(\xi_{ij}(L_k)) \to C - F^{-1}(p) = z_j(k)$$

**Theorem 5.2** Under Assumption 5.2, the first moment of the estimate Eq. (5.6) in the step (3) of ACA can be given by

$$|E(\hat{x}_{ij}(L_k)) - z_j(k)| = O\left(\frac{1}{k^{\beta/2}}\right)$$

**Proof** For the estimate Eq. (5.6), we have by [[10], Theorem 4.7]

$$E(\hat{x}_{ij}(L_k) - z_j(k))^2 = O\left(\frac{1}{L_k}\right) = O\left(\frac{1}{k^\beta}\right)$$

In addition

$$E^2(\hat{x}_{ij}(L_k) - z_j(k)) \leq E(\hat{x}_{ij}(L_k) - z_j(k))^2$$

we have

$$|E(\hat{x}_{ij}(L_k) - z_j(k))| = O\left(\frac{1}{k^{\beta/2}}\right) \tag{5.16}$$

### 5.3.2 Transportation Design

By Eq. (5.12) and Eq. (5.13), the state is updated in two ways:

(1) If there is an agent who doesn't meet condition Eq. (5.8), the states updating will be Eq. (5.12), which can be written in a vector form:

$$z(k+1) = z(k) \tag{5.17}$$

(2) If all the agents satisfy condition Eq. (5.8), the states updating will be Eq. (5.13), which can be written in another form:

$$z_i(k+1) = z_i(k) - \frac{1}{k} \sum_{(j,i) \in G_I} (z_i(k) - z_j(k) + z_j(k) - \hat{x}_{ij}(L_k)) +$$

$$\frac{1}{k} \sum_{(i,j) \in G_I} (z_j(k) - z_i(k) + z_i(k) - \hat{x}_{ji}(L_k))$$

Denote $\hat{x}_{ij}(L_k) - z_j(k) = \varepsilon_{ij}(k)$ and $N_i = \{j, (j,i) \in G_I\} \cup \{j, (i,j) \in G_I\}$, the state updating Eq. (5.13) can be written as

$$z_i(k+1) = z_i(k) - \frac{1}{k} \sum_{j \in N_i} (z_i(k) - z_j(k)) + \frac{1}{k} \sum_{(j,i) \in G_I} \varepsilon_{ij}(k) - \frac{1}{k} \sum_{(i,j) \in G_I} \varepsilon_{ji}(k)$$

Let $\varepsilon(k)$ be the $l_s$ dimensional vector that contains all $\varepsilon_{ij}(k)$ in a selected order. The above equation can be written in a vector form:

$$z(k+1) = z(k) - \frac{M}{k}z(k) + \frac{W}{k}\varepsilon(k) \tag{5.18}$$

where $M$ is the Laplacian Matrix of the undirected graph with the same nodes and links as $G_I$, and $W$ is $n \times l_s$ matrix whose $v$th column are of all zeros except for "1" at the $i$th position and "−1" at the $j$th position if the $v$th element of $\varepsilon(k)$ is $\varepsilon_{ij}(k)$. Taking Fig. 5.1 for example, $n=5$, $l_s=6$, we have

$$\varepsilon(k) = \begin{bmatrix} \varepsilon_{12}(k) \\ \varepsilon_{23}(k) \\ \varepsilon_{34}(k) \\ \varepsilon_{45}(k) \\ \varepsilon_{51}(k) \\ \varepsilon_{14}(k) \end{bmatrix}, \quad M = \begin{bmatrix} 3 & -1 & 0 & -1 & -1 \\ -1 & 2 & -1 & 0 & 0 \\ 0 & -1 & 2 & -1 & 0 \\ -1 & 0 & -1 & 3 & -1 \\ -1 & 0 & 0 & -1 & 2 \end{bmatrix}$$

## 5.3 Properties of the Algorithm

$$W = \begin{bmatrix} 1 & 0 & 0 & 0 & -1 & 1 \\ -1 & 1 & 0 & 0 & 0 & 0 \\ 0 & -1 & 1 & 0 & 0 & 0 \\ 0 & 0 & -1 & 1 & 0 & -1 \\ 0 & 0 & 0 & -1 & 1 & 0 \end{bmatrix}$$

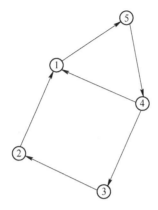

Fig. 5.1  A directed network with five agents

Considering the above two ways together, we can get a unified state updating expression by Eq. (5.17) and Eq. (5.18)

$$z(k+1) = z(k) - \left(\frac{M}{k}z(k) - \frac{W}{k}\varepsilon(k)\right) I_{\{\bigcap_{i=1}^{n} |x_i^1(t_{k+1})| < B + \sum_{h=1}^{k+1} \frac{1}{h^{1+\gamma}}\}} \quad (5.19)$$

**Remark 5.3**  Since $\mathbb{1}_n^T M = 0$, $\mathbb{1}_n^T W = 0$, we have by (5.19)

$$\mathbb{1}_n^T z(k+1) = \mathbb{1}_n^T z(k) = \bar{x}n$$

which implies that the sum of the agents' states is a constant.

### 5.3.2.1 Convergence

Denote $z_i^1(k) = x_i^1(t_k)$. Then, the Eq. (5.19) can be written as follows:

$$z(k+1) = \left(I - \frac{M}{k}\right)z(k) + \frac{W}{k}\varepsilon(k) + \left(\frac{M}{k}z(k) - \frac{M}{k}\varepsilon(k)\right) \cdot I_{\{\bigcup_{i=1}^{n} |z_i^1(k+1)| \geqslant B + \sum_{h=1}^{k+1} \frac{1}{h^{1+\gamma}}\}}$$

(5.20)

Denote

$$R(k) = (Mz(k) - W\varepsilon(k)) I_{\{\bigcup_{i=1}^{n} |z_i^1(k+1)| \geqslant B + \sum_{h=1}^{k+1} \frac{1}{h^{1+\gamma}}\}} \quad (5.21)$$

Subtracting $\bar{x}\mathbb{1}_n$ to the Eq. (5.20) results in

$$z(k+1) - \bar{x}\mathbb{1}_n$$
$$= \left(I_n - \frac{1}{k}M\right)(z(k) - \bar{x}\mathbb{1}_n) - \frac{1}{k}\bar{x}M\mathbb{1}_n + \frac{1}{k}W\varepsilon(k) + \frac{1}{k}R(k)$$
$$= \left(I_n - \frac{1}{k}M\right)(z(k) - \bar{x}\mathbb{1}_n) + \frac{1}{k}W\varepsilon(k) + \frac{1}{k}R(k)$$

Denote $e(k) = z(k) - \bar{x}\mathbb{1}_n$. We have

$$e(k+1) = \left(I_n - \frac{1}{k}M\right)e(k) + \frac{1}{k}W\varepsilon(k) + \frac{1}{k}R(k) \tag{5.22}$$

Then,

$$\mathbb{1}_n^T e(k+1) = \mathbb{1}_n^T e(k) = \mathbb{1}_n^T z(k) - \mathbb{1}_n^T \bar{x}\mathbb{1}_n = 0 \tag{5.23}$$

Decompose $M$ as

$$M = \begin{bmatrix} M_{11} & M_{12} \\ M_{21} & M_{22} \end{bmatrix}$$

where $M_{11} \in R^{(n-1)\times(n-1)}$, $M_{12} \in R^{(n-1)\times 1}$, $M_{21} \in R^{(n-1)\times 1}$, $M_{22} \in R$. Then

$$I_n - \frac{1}{k}M = \begin{bmatrix} I_{n-1} - \frac{1}{k}M_{11} & -\frac{1}{k}M_{12} \\ -\frac{1}{k}M_{21} & 1 - \frac{1}{k}M_{22} \end{bmatrix}$$

Accordingly, we also decompose $e(k)$, $W$ and $R(k)$ as

$$e(k) = \begin{bmatrix} \tilde{e}(k) \\ e_n(k) \end{bmatrix}, \quad W = \begin{bmatrix} \tilde{W} \\ W_n \end{bmatrix} \text{ and } R(k) = \begin{bmatrix} \tilde{R}(k) \\ R_n(k) \end{bmatrix}$$

Eq. (5.22) can be divided as:

$$\tilde{e}(k+1) = \left(I_{n-1} - \frac{1}{k}M_{11}\right)\tilde{e}(k) - \frac{1}{k}M_{12}e_n(k) + \frac{1}{k}\tilde{W}\varepsilon(k) + \frac{1}{k}\tilde{R}(k)$$
$$\tag{5.24}$$

and

$$e_n(k+1) = -\frac{1}{k}M_{21}\tilde{e}(k) + \left(1 - \frac{1}{k}M_{22}\right)e_n(k) + \frac{1}{k}W_n\varepsilon(k) + \frac{1}{k}R_n(k)$$
$$\tag{5.25}$$

By Eq. (5.23), we have $\mathbb{1}_{n-1}^T \tilde{e}(k) + e_n(k) = 0$. Hence we just need to calculate $\tilde{e}(k+1)$ as follows:

$$\tilde{e}(k+1)$$
$$= \left(I_{n-1} - \frac{1}{k}M_{11}\right)\tilde{e}(k) + \frac{1}{k}M_{12}\mathbb{1}_{n-1}^T \tilde{e}(k) + \frac{1}{k}\widetilde{W}\varepsilon(k) + \frac{1}{k}\tilde{R}(k)$$
$$= \left(I_{n-1} - \frac{1}{k}\widetilde{M}_{11}\right)\tilde{e}(k) + \frac{1}{k}\widetilde{W}\varepsilon(k) + \frac{1}{k}\tilde{R}(k) \qquad (5.26)$$

where $\widetilde{M}_{11} = M_{11} - M_{12}\mathbb{1}_{n-1}^T$.

To prove the convergence, we first give two lemmas as follows.

**Lemma 5.1** Under Assumption 5.1, the eigenvalues of $\widetilde{M}_{11}$ are positive, which are the non-zero eigenvalues of $M$.

**Proof** Under Assumption 5.1, the undirected graph $G_M$ with the same nodes and links as $G_I$ is connected. The matrix $M$ is the Laplacian matrix of $G_M$. By Theorem 2.8, we have

$$0 = \lambda_1(M) < \lambda_2(M) \leq \lambda_3(M) \leq \cdots \leq \lambda_n(M) \qquad (5.27)$$

where $\lambda_i(M)$, $i=1, \cdots, n$ are the eigenvalues of Laplacian matrix $M$.

Since
$$M = \begin{bmatrix} M_{11} & M_{12} \\ M_{21} & M_{22} \end{bmatrix}$$

and
$$\mathbb{1}_n^T M = 0$$

We have
$$|\lambda I - M| = \begin{vmatrix} \lambda I_{n-1} - M_{11} & -M_{12} \\ -M_{21} & \lambda - M_{22} \end{vmatrix}$$

Add the left-multiplying the first row of the right side of the above equation by $\mathbb{1}_{n-1}^T$ to the second row, then

$$|\lambda I - M| = \begin{vmatrix} \lambda I_{n-1} - M_{11} & -M_{12} \\ \lambda \mathbb{1}_{n-1}^T & \lambda \end{vmatrix}$$
$$= \begin{vmatrix} \lambda I_{n-1} - M_{11} + M_{12}\mathbb{1}_{n-1}^T & -M_{12} \\ 0 & \lambda \end{vmatrix}$$
$$= \lambda |\lambda I_{n-1} - (M_{11} - M_{12}\mathbb{1}_{n-1}^T)|$$
$$= \lambda |\lambda I_{n-1} - \widetilde{M}_{11}|$$

Thus, the eigenvalues of $\widetilde{M}_{11}$ are the eigenvalues of $M$. By Eq. (5.27), we can get the lemma.

**Lemma 5.2** Under Assumption 5.2, we have the assertion on $R(k)$ defined in Eq. (5.21) as follows,

$$E(R^T(k)R(k)) = O\left(\frac{1}{k^{\beta-2\gamma}}\right)$$

**Proof** Since $R(k)$ is defined as

$$R(k) = (Mz(k) - W\varepsilon(k))I\left\{\bigcup_{i=1}^{n} |z_i^1(k+1)| \geq B + \sum_{h=1}^{k+1}\frac{1}{h^{1+\gamma}}\right\}$$

Then

$$R^T(k)R(k) = z^T(k)M^2z(k)I\left\{\bigcup_{i=1}^{n} |z_i^1(k+1)| \geq B + \sum_{h=1}^{k+1}\frac{1}{h^{1+\gamma}}\right\} -$$

$$2z^T(k)MW\varepsilon(k)I\left\{\bigcup_{i=1}^{n} |z_i^1(k+1)| \geq B + \sum_{h=1}^{k+1}\frac{1}{h^{1+\gamma}}\right\} +$$

$$\varepsilon^T(k)W^TW\varepsilon(k)I\left\{\bigcup_{i=1}^{n} |z_i^1(k+1)| \geq B + \sum_{h=1}^{k+1}\frac{1}{h^{1+\gamma}}\right\}$$

Taking expectation on each item of the right side of the above equation, since $z(k)$ is bounded, we have

$$E\left(z^T(k)M^2z(k)I\left\{\bigcup_{i=1}^{n} |z_i^1(k+1)| \geq B + \sum_{h=1}^{k+1}\frac{1}{h^{1+\gamma}}\right\}\right) = O\left(P\left\{\bigcup_{i=1}^{n} |z_i^1(k+1)| \geq B + \sum_{h=1}^{k+1}\frac{1}{h^{1+\gamma}}\right\}\right)$$

$$E\left(z^T(k)MW\varepsilon(k)I\left\{\bigcup_{i=1}^{n} |z_i^1(k+1)| \geq B + \sum_{h=1}^{k+1}\frac{1}{h^{1+\gamma}}\right\}\right)$$
$$= O\left(\sqrt{E\varepsilon^T(k)\varepsilon(k)}\sqrt{P\left\{\bigcup_{i=1}^{n} |z_i^1(k+1)| \geq B + \sum_{h=1}^{k+1}\frac{1}{h^{1+\gamma}}\right\}}\right)$$

and

$$E\left(\varepsilon^T(k)W^TW\varepsilon(k)I\left\{\bigcup_{i=1}^{n} |z_i^1(k+1)| \geq B + \sum_{h=1}^{k+1}\frac{1}{h^{1+\gamma}}\right\}\right) = O(E\varepsilon^T(k)\varepsilon(k))$$

Now, we pay attention to

$$P\left\{\bigcup_{i=1}^{n} |z_i^1(k+1)| \geq B + \sum_{h=1}^{k+1}\frac{1}{h^{1+\gamma}}\right\}$$

$$P\left\{\bigcup_{i=1}^{n} |z_i^1(k+1)| \geq B + \sum_{h=1}^{k+1}\frac{1}{h^{1+\gamma}}\right\}$$

$$\leq \sum_{i=1}^{n} P\left\{|z_i^1(k+1)| \geq B + \sum_{h=1}^{k+1}\frac{1}{h^{1+\gamma}}\right\}$$

$$\leq O\left(P\left\{\left|z_i(k) - \frac{1}{k}\sum_{j \in N_i}(z_i(k) - z_j(k))\right| + \right.\right.$$

$$\frac{1}{k}\sum_{(j,i)\in G_I}\varepsilon_{ij}(k) - \frac{1}{k}\sum_{(i,j)\in G_I}\varepsilon_{ji}(k) \Big| \geq B + \sum_{h=1}^{k+1}\frac{1}{h^{1+\gamma}}\Big\}\Big)$$

$$\leq O\Big(P\Big\{\frac{l_s}{k}|\varepsilon_{ij}(k)| \geq B + \sum_{h=1}^{k+1}\frac{1}{h^{1+\gamma}} - \Big|z_i(k) - \frac{1}{k}\sum_{j\in N_i}(z_i(k) - z_j(k))\Big|\Big\}\Big)$$

Since

$$z_i(k) - \frac{1}{k}\sum_{j\in N_i}(z_i(k) - z_j(k)) = \Big(1 - \frac{|N_i|}{k}\Big)z_i(k) + \frac{1}{k}\sum_{j\in N_i}z_j(k)$$

We have

$$\Big(1 - \frac{|N_i|}{k}\Big)z_i(k) + \frac{|N_i|}{k}\min_{j\in N_i}z_j(k) \leq z_i(k) - \frac{1}{k}\sum_{j\in N_i}(z_i(k) - z_j(k))$$

$$\leq \Big(1 - \frac{|N_i|}{k}\Big)z_i(k) + \frac{|N_i|}{k}\max_{j\in N_i}z_j(k)$$

where $|N_i|$ means the number of the elements in the set $N_i$. Also for any $d_1$, $d_2$, $0 < \rho < 1$, $\min\{d_1, d_2\} \leq (1-\rho)d_1 + \rho d_2 \leq \max\{d_1, d_2\}$, thus

$$\min\{z_i(k), \min_{j\in N_i}z_j(k)\} \leq z_i(k) - \frac{1}{k}\sum_{j\in N_i}(z_i(k) - z_j(k))$$

$$\leq \max\{z_i(k), \max_{j\in N_i}z_j(k)\}, \quad \text{as } k > n$$

Besides, $|z_i(k)| < B + \sum_{h=1}^{k}\frac{1}{h^{1+\gamma}}$ for $i = 1, \cdots, n$. Then, we have

$$\Big|z_i(k) - \frac{1}{k}\sum_{j\in N_i}(z_i(k) - z_j(k))\Big| < B + \sum_{h=1}^{k}\frac{1}{h^{1+\gamma}}, \quad \text{as } k > n$$

Hence,

$$P\Big\{\frac{l_s}{k}|\varepsilon_{ij}(k)| \geq B + \sum_{h=1}^{k+1}\frac{1}{h^{1+\gamma}} - \Big|z_i(k) - \frac{1}{k}\sum_{j\in N_i}(z_i(k) - z_j(k))\Big|\Big\}$$

$$\leq P\Big\{\frac{l_s}{k}|\varepsilon_{ij}(k)| \geq \frac{1}{(k+1)^{1+\gamma}}\Big\}$$

$$\leq \frac{E(|\varepsilon_{ij}(k)|)^2}{\frac{k^2}{l_s^2(k+1)^{2+2\gamma}}} = O\Big(\frac{1}{k^{\beta-2\gamma}}\Big)$$

Taking expectation on $R^T(k)R(k)$, we can get the lemma by

$$E(R^T(k)R(k))$$
$$= O\left(\frac{1}{k^{\beta-2\gamma}}\right) + O\left(\sqrt{\frac{1}{k^\beta}\frac{1}{k^{\beta-2\gamma}}}\right) + O\left(\frac{1}{k^\beta}\right)$$
$$= O\left(\frac{1}{k^{\beta-2\gamma}}\right)$$

**Theorem 5.3** Under Assumptions 5.1 and 5.2, we can get that the skipping state $z(k)$ updated by Eq. (5.19) can converge to the average, that is to say,
$$z(k) \to \bar{x}\mathbb{1}_n, \text{ w.p.1 as } k \to \infty$$

**Proof** To prove $z(k) \to \bar{x}\mathbb{1}_n$ w.p.1, we need to prove $e(k) \to 0$ w.p.1. Since $\mathbb{1}_n^T e(k) = \mathbb{1}_{n-1}^T \tilde{e}(k) + e_n(k) = 0$ by Eq. (5.23), we only need to prove $\tilde{e}(k) \to 0$ w.p.1.

Let $Q_k = I_{n-1} - \frac{1}{k}\tilde{M}_{11}$, we have by Eq. (5.26)

$$\tilde{e}(k+1) = Q_k \tilde{e}(k) + \frac{1}{k}\tilde{W}\varepsilon(k) + \frac{1}{k}\tilde{R}(k) \tag{5.28}$$

Thus
$$\tilde{e}(k+1)^T \tilde{e}(k+1)$$
$$= \tilde{e}(k)^T Q_k^T Q_k \tilde{e}(k) + \frac{2}{k}\tilde{e}(k)^T Q_k^T (\tilde{W}\varepsilon(k) + \tilde{R}(k)) +$$
$$\frac{1}{k^2}(\tilde{W}\varepsilon(k) + \tilde{R}(k))^T(\tilde{W}\varepsilon(k) + \tilde{R}(k)) \tag{5.29}$$

By Lemma 5.1, we have
$$0 < \lambda(Q_k) < 1 \text{ as } k \text{ is large enough}$$

Thus
$$\tilde{e}(k)^T Q_k^T Q_k \tilde{e}(k) \leq \lambda_{max}(Q_k^T Q_k)\tilde{e}(k)^T \tilde{e}(k) \leq \tilde{e}(k)^T \tilde{e}(k)$$

Together with Eq. (5.29), we have
$$\tilde{e}(k+1)^T \tilde{e}(k+1)$$
$$\leq \tilde{e}(k)^T \tilde{e}(k) + \frac{2}{k}\tilde{e}(k)^T Q_k^T(\tilde{W}\varepsilon(k) + \tilde{R}(k)) +$$
$$\frac{1}{k^2}(\tilde{W}\varepsilon(k) + \tilde{R}(k))^T(\tilde{W}\varepsilon(k) + \tilde{R}(k)) \tag{5.30}$$

Denote $\mathcal{F}_k = \sigma\{\tilde{e}(k), \tilde{e}(k-1), \cdots, \tilde{e}(1)\}$. Then $\tilde{e}(k)$ is $\mathcal{F}_k$-measurable, we have
$$E(\tilde{e}(k)^T \tilde{e}(k) | \mathcal{F}_k) = \tilde{e}(k)^T \tilde{e}(k), \text{ w.p.1 as } k \to \infty$$

## 5.3 Properties of the Algorithm

Taking conditional expectation on inequality Eq. (5.30):

$$E(\tilde{e}(k+1)^T\tilde{e}(k+1)\mid\mathcal{F}_k)$$
$$\leqslant \tilde{e}(k)^T\tilde{e}(k) + \frac{2}{k}E(\tilde{e}(k)^T Q_k^T(\widetilde{W}\varepsilon(k) + \tilde{R}(k))\mid\mathcal{F}_k) +$$
$$\frac{1}{k^2}E((\widetilde{W}\varepsilon(k) + \tilde{R}(k))^T(\widetilde{W}\varepsilon(k) + \tilde{R}(k))\mid\mathcal{F}_k)$$

Due to the property of conditional expectation, we can obtain

$$E[E((\widetilde{W}\varepsilon(k) + \tilde{R}(k))^T(\widetilde{W}\varepsilon(k) + \tilde{R}(k))\mid\mathcal{F}_k)]$$
$$= E(\widetilde{W}\varepsilon(k) + \tilde{R}(k))^T(\widetilde{W}\varepsilon(k) + \tilde{R}(k))$$

By Holder inequality,

$$E(\widetilde{W}\varepsilon(k) + \tilde{R}(k))^T(\widetilde{W}\varepsilon(k) + \tilde{R}(k))$$
$$\leqslant E(\varepsilon(k)^T\widetilde{W}^T\widetilde{W}\varepsilon(k)) + 2\sqrt{E(\varepsilon(k)^T\widetilde{W}^T\widetilde{W}\varepsilon(k))}\sqrt{E(\tilde{R}(k)^T\tilde{R}(k))} +$$
$$E(\tilde{R}(k)^T\tilde{R}(k))$$

By Theorem 5.1, we can obtain

$$E(\varepsilon(k)^T\widetilde{W}^T\widetilde{W}\varepsilon(k)) = O\left(\frac{1}{k^\beta}\right)$$

Since $\mathbb{1}_n^T R(k) = \mathbb{1}_{n-1}^T \tilde{R}(k) + R_n(k) = 0$, we have

$$E(R^T(k)R(k)) = E(\tilde{R}^T(k)\tilde{R}(k) + R_n^T(k)R_n(k))$$
$$= E(\tilde{R}^T(k)(I_{n-1} + \mathbb{1}_{n-1}\mathbb{1}_{n-1}^T)\tilde{R}(k))$$

By Lemma 5.2, we have

$$E(\tilde{R}^T(k)\tilde{R}(k)) = O(E(R^T(k)R(k))) = O\left(\frac{1}{k^{\beta-2\gamma}}\right)$$

Hence

$$E(\widetilde{W}\varepsilon(k) + \tilde{R}(k))^T(\widetilde{W}\varepsilon(k) + \tilde{R}(k))$$
$$= O\left(\frac{1}{k^\beta}\right) + O\left(\sqrt{\frac{1}{k^\beta}\frac{1}{k^{\beta-2\gamma}}}\right) + O\left(\frac{1}{k^{\beta-2\gamma}}\right) = O\left(\frac{1}{k^{\beta-2\gamma}}\right)$$

Analogously

$$E[E(\tilde{e}(k)^T Q_k^T(\widetilde{W}\varepsilon(k) + \tilde{R}(k))\mid\mathcal{F}_k)]$$
$$= E(\tilde{e}(k)^T Q_k^T(\widetilde{W}\varepsilon(k) + \tilde{R}(k)))$$
$$\leqslant \sqrt{E(\tilde{e}(k)^T Q_k^T Q_k \tilde{e}(k))}\sqrt{E(\widetilde{W}\varepsilon(k) + \tilde{R}(k))^T(\widetilde{W}\varepsilon(k) + \tilde{R}(k))}$$

Since the ACA ensures the boundedness of $z_i(k)$, it follows that $E(\tilde{e}(k)^T Q_k^T Q_k \tilde{e}(k))$ is bounded. Then

$$\frac{2}{k} E[E(\tilde{e}(k)^T Q_k^T (\widetilde{W}\varepsilon(k) + \tilde{R}(k)) | \mathcal{F}_k)] +$$

$$\frac{2}{k^2} E[E((\widetilde{W}\varepsilon(k) + \tilde{R}(k))^T (\widetilde{W}\varepsilon(k) + \tilde{R}(k)) | \mathcal{F}_k)]$$

$$\leq O\left(\frac{1}{k^{1+\beta/2-\gamma}}\right) + O\left(\frac{1}{k^{2+\beta-2\gamma}}\right) = O\left(\frac{1}{k^{1+\beta/2-\gamma}}\right)$$

Denote $\delta = \beta/2 - \gamma$. Then $\delta > 0$ since $\gamma < \beta/2$. Thus

$$\sum_{k=1}^{\infty} O\left(\frac{1}{k^{1+\delta}}\right) < \infty$$

According to [[50], Lemma 1.2.2], we obtain

$$\tilde{e}(k+1)^T \tilde{e}(k+1) \to a, \text{ w.p.1 } \text{ as } k \to \infty \qquad (5.31)$$

The next step is to prove that $\tilde{e}(k) \to 0$. Taking expectations on both sides of Eq. (5.29), we have

$$E(\tilde{e}(k+1)^T \tilde{e}(k+1)) = E\tilde{e}(k)^T Q_k^T Q_k \tilde{e}(k) + O\left(\frac{1}{k^{1+\delta}}\right) \qquad (5.32)$$

Taking $e(k) = Q_{k-1} \tilde{e}(k-1) + \frac{1}{k-1} \widetilde{W}\varepsilon(k-1) + \frac{1}{k-1} \tilde{R}(k-1)$ by Eq. (5.28) into Eq. (5.32), we have

$$E(\tilde{e}(k+1)^T \tilde{e}(k+1))$$

$$= E(\tilde{e}(k+1)^T Q_{k-1}^T Q_k^T Q_k Q_{k-1} \tilde{e}(k-1)) + O\left(\frac{1}{(k-1)^{1+\delta}} Q_k^T Q_k\right) + O\left(\frac{1}{k^{1+\delta}}\right)$$

Analogously, we can get the following equation

$$E(\tilde{e}(k+1)^T \tilde{e}(k+1))$$

$$= E(\tilde{e}(1)^T Q_1^T Q_2^T \cdots Q_k^T Q_k \cdots Q_2 Q_1 \tilde{e}(1)) + O\left(\frac{1}{k^{1+\delta}}\right) +$$

$$O\left(\sum_{l=1}^{k-1} \frac{1}{l^{1+\delta}} \mathbb{1}_{n-1}^T Q_{l+1}^T Q_{l+2}^T \cdots Q_k^T Q_k \cdots Q_{l+2} Q_{l+1} \mathbb{1}_{n-1}\right)$$

$$= E(\| Q_k \cdots Q_2 Q_1 \tilde{e}(1) \|_2^2) + O\left(\sum_{l=1}^{k-1} \frac{1}{l^{1+\delta}} \| Q_k \cdots Q_{l+2} Q_{l+1} \mathbb{1}_{n-1} \|_2^2\right) + O\left(\frac{1}{k^{1+\delta}}\right)$$

## 5.3 Properties of the Algorithm

Noticing that there exists $T$ such that

$$T^{-1}\widetilde{M}_{11}T = \Lambda = \mathrm{diag}(\lambda_1, \lambda_2, \cdots, \lambda_{n-1})$$

where $\lambda_i$ is the eigenvalue of $\widetilde{M}_{11}$, we have

$$\|Q_k \cdots Q_{l+2} Q_{l+1}\|_2^2 = \left\|T\left(I - \frac{1}{k}\Lambda\right)\cdots\left(I - \frac{1}{l+2}\Lambda\right)\left(I - \frac{1}{l+1}\Lambda\right)T^{-1}\right\|_2^2$$

$$\leqslant \|T\|_2^2 \|T^{-1}\|_2^2 \prod_{i=l+1}^{k}\left(1 - \frac{1}{i}\underline{\lambda}\right)^2, \quad \text{as } l \geqslant \overline{\lambda}$$

where $\underline{\lambda} = \lambda_{\min}(\widetilde{M}_{11})$, $\overline{\lambda} = \lambda_{\max}(\widetilde{M}_{11})$. Therefore,

$$E(\tilde{e}(k+1)^T \tilde{e}(k+1))$$

$$= O\left(\prod_{i=\overline{\lambda}+1}^{k}\left(1 - \frac{1}{i}\underline{\lambda}\right)^2\right) + O\left(\sum_{l=\overline{\lambda}}^{k-1}\frac{1}{l^{1+\delta}}\prod_{i=l+1}^{k}\left(1 - \frac{1}{i}\underline{\lambda}\right)^2\right) + O\left(\frac{1}{k^{1+\delta}}\right)$$

Denote $\widetilde{V}_k = E(\tilde{e}(k)^T \tilde{e}(k))$. According to Theorem 2.10, we have

$$\widetilde{V}_{k+1} = \begin{cases} O\left(\dfrac{1}{k^\delta}\right) & \delta < 2\underline{\lambda} \\ O\left(\dfrac{\ln k}{k^{2\underline{\lambda}}}\right) & \delta = 2\underline{\lambda} \\ O\left(\dfrac{1}{k^{2\underline{\lambda}}}\right) & \delta > 2\underline{\lambda} \end{cases} \quad (5.33)$$

Sequentially,

$$E(\tilde{e}(k+1)^T \tilde{e}(k+1)) = \widetilde{V}_{k+1} \to 0$$

Together with Eq. (5.31), we have

$$\tilde{e}(k) \to 0, \text{ w.p. } 1$$

Hence, the theorem is proved.

As we mentioned above, $x(t_k) = z(k)$. According to theorem 5.3

$$x(t_k) \to \bar{x}\, \mathbb{1}_n, \text{ w.p. } 1 \quad \text{as} \quad k \to \infty$$

For $\forall t \in N$, there exists a constant $k$ such that $t \in [t_k, t_{k+1})$. Thus $x(t) = x(t_k)$, and

$$k \to \infty \Leftrightarrow t \to \infty$$

Then,

$$\lim_{t \to \infty} x(t) = \lim_{k \to \infty} x(t_k) = \bar{x}\, \mathbb{1}_n \quad (5.34)$$

Hence, we can get the following theorem.

**Theorem 5.4** Under Assumptions 5.1 and 5.2, the multi-agent system can achieve average consensus, that is to say,

$$x(t) \to \bar{x} \mathbb{1}_n, \text{ w. p. 1} \quad \text{as} \quad t \to \infty$$

#### 5.3.2.2 Convergence Rate

The mean square convergence rate of $x(t_k)$, $k = 1, 2 \cdots$, is given in expression (5.33), from which we can see that the larger $\delta$, the higher the convergence rate. When the distributed graph $G_l$ of the system is given, the parameter $\underline{\lambda} = \lambda_{\min}(\widetilde{M}_{11})$ is a constant. A larger $\beta$ can be designed such that $\delta = \beta/2 - \gamma > 2\underline{\lambda}$, which makes the skipping state $z(k)$ converge to $\bar{x} \mathbb{1}_n$ with the convergence rate $O\left(\dfrac{1}{k^{2\underline{\lambda}}}\right)$. This is consistent with the fact that the more time the agents spend estimating, the faster the state $z(k)$ given by iteration (5.18) converges.

However, expression (5.33) is the convergence rate of skip state $x(t_k)$, $k = 1, 2, \cdots$, not the continuous state $x(t)$, $t = 1, 2, \cdots$. When $\beta$ is large enough, $z(k)$ can converge fast, whereas the state $x(t)$, $t = 1, 2, \cdots$ may converge slowly due to long time keeping the states unchanged for estimation.

Next, we will discuss the convergence rate of $x(t)$ and the result is given in the following theorem.

**Theorem 5.5** Under Assumptions 5.1 and 5.2, the consensus speed can be given as follows:

$$E((x(t) - \bar{x}\mathbb{1}_n)^T(x(t) - \bar{x}\mathbb{1}_n)) = \begin{cases} O\left(\dfrac{1}{(1+\beta)\sqrt{t^{\beta/2-\gamma}}}\right) & \beta/2 - \gamma < 2\underline{\lambda} \\ O\left(\dfrac{\ln t}{(1+\beta)\sqrt{t^{2\underline{\lambda}}}}\right) & \beta/2 - \gamma = 2\underline{\lambda} \\ O\left(\dfrac{1}{(1+\beta)\sqrt{t^{2\underline{\lambda}}}}\right) & \beta/2 - \gamma > 2\underline{\lambda} \end{cases}$$

where $\beta$ is the parameter characterize the estimation time, $\gamma$ is a constant satisfies $0 < \gamma < \beta/2$, $\underline{\lambda}$ is the smallest non-negative eigenvalue of Laplacian Matrix $M$.

**Proof** Denote $V_k = E(e^T(k)e(k))$. Then

## 5.3 Properties of the Algorithm

$$V_k = E(e^T(k)e(k))$$
$$= E(\tilde{e}^T(k)\tilde{e}(k) + e_n^T(k)e_n(k))$$
$$= E(\tilde{e}^T(k)\tilde{e}(k)) + E((\mathbb{1}_{n-1}^T\tilde{e}(k))^T(\mathbb{1}_{n-1}^T\tilde{e}(k)))$$

Also

$$E((\mathbb{1}_{n-1}^T\tilde{e}(k))^T(\mathbb{1}_{n-1}^T\tilde{e}(k))) = E((\tilde{e}(k))^T \mathbb{1}_{n-1}\mathbb{1}_{n-1}^T\tilde{e}(k))$$
$$\leq \lambda_{max}(\mathbb{1}_{n-1}\mathbb{1}_{n-1}^T)E(\tilde{e}^T(k)\tilde{e}(k))$$
$$\leq \lambda_{max}(\mathbb{1}_{n-1}\mathbb{1}_{n-1}^T)\tilde{V}_k$$

We have

$$V_k \leq (1 + \lambda_{max}(\mathbb{1}_{n-1}\mathbb{1}_{n-1}^T))\tilde{V}_k$$

Thus

$$V_k = O(\tilde{V}_k) \tag{5.35}$$

For $\forall t \in N$, there exists a constant $k$ such that $t \in [t_k, t_{k+1})$. Then

$$O(t_k) \leq O(t) \leq O(t_{k+1}), \quad x(t) = x(t_k), \quad k \to \infty \Leftrightarrow t \to \infty$$

As a matter of fact

$$t_k = t_{k-1} + L_{k-1} = \sum_{l=1}^{k-1} L_l = \sum_{l=1}^{k-1} l_\beta = O(k^{\beta+1})$$

So $O(t_k) = O(t_{k+1}) = O(k^{\beta+1})$, this implies

$$O(k) = O(\sqrt[(1+\beta)]{t}), \quad \text{if} \quad t \in [t_k, t_{k+1})$$

Taking this into Eq. (5.33), we can get the theorem.

**Remark 5.4** If $\beta/2 - \gamma < 2\underline{\lambda}$, i.e., $\beta < 4\underline{\lambda} + \gamma$, then

$$V_k = O\left(\frac{1}{\sqrt[(1+\beta)]{t^{\beta/2-\gamma}}}\right) = O\left(\frac{1}{t^{\frac{\beta-2\gamma}{2(1+\beta)}}}\right)$$

which implies that the larger $\beta$, the faster the convergence rate. But $\beta$ cannot be too large because it should satisfy $\beta < 4\underline{\lambda} + \gamma$. If $\beta > 4\underline{\lambda} + \gamma$, then

$$V_k = O\left(\frac{1}{\sqrt[(1+\beta)]{t^{2\underline{\lambda}}}}\right) = O\left(\frac{1}{t^{\frac{2\underline{\lambda}}{1+\beta}}}\right)$$

which implies that the smaller $\beta$ the better. But $\beta$ cannot be too small because it should be bigger than $4\underline{\lambda} + \gamma$. From the above analysis, it's concluded that $\beta$ can neither be too big nor too small for taking the actual convergence rate into consideration.

## 5.4 Numerical Simulation

In simulation, the directed network with five agents as Fig. 5.1 is considered. The states of the agents are updated by $x_i(t+1) = x_i(t) + u_i(t)$, $i=1, \cdots, n$. Each agent can only get binary-valued information from its neighbors by (5.3) with Gaussian noises $N(0, 5)$ and the threshold $C=3$. The control $u_i(t)$ is realized by a quantity of transportation, which is designed based on the binary-valued information.

Let the initial state $x(0) = [0, 1, 3, 8, 13]^T$. The average value of initial states is 5. By ACA in Section 5.2.1, each agent estimates its neighbors states for a holding time $L_k$, and then designs the quantity of transportation, which leads the states of the system being updated by (5.12) or (5.13). In the simulation, we use the holding time $L_k = k^\beta$ with $\beta = 0.2$, $\beta = 1$, $\beta = 2$, $\beta = 4$ respectively to estimate the states. Then, the control is designed and the states are updated. Fig. 5.2 shows the trajectories of the skipping states $z_i(k)$, $i=1, \cdots, 5$. Fig. 5.3 shows the continuous states $x_i(t)$, $i=1, \cdots, 5$. Fig. 5.4 is the variance of the states $x_i(t)$.

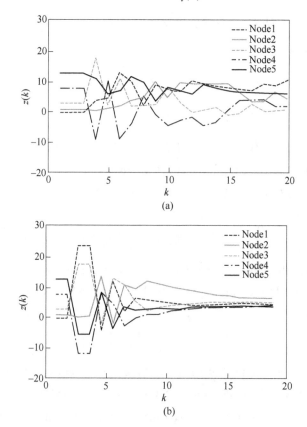

(a)

(b)

5.4 Numerical Simulation

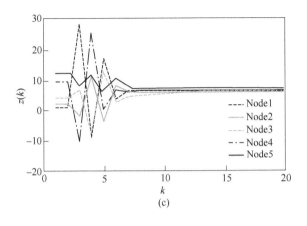

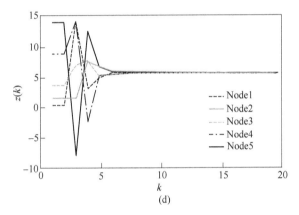

Fig. 5.2 Trajectories of skipping states $z(k)$ with different $\beta$
(a) $\beta=0.2$; (b) $\beta=1$; (c) $\beta=2$; (d) $\beta=4$

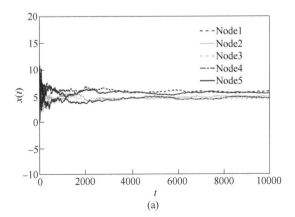

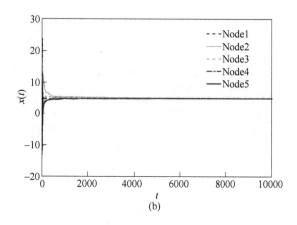

(b)

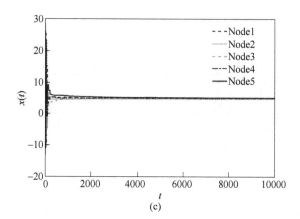

(c)

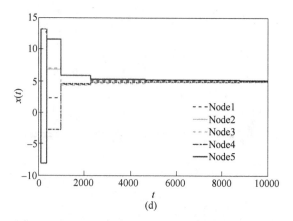

(d)

Fig. 5.3　Trajectories of states $x(t)$ with different $\beta$
(a) $\beta=0.2$; (b) $\beta=1$; (c) $\beta=2$; (d) $\beta=4$

## 5.4 Numerical Simulation

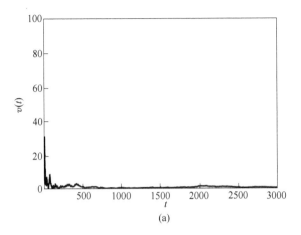

(a)

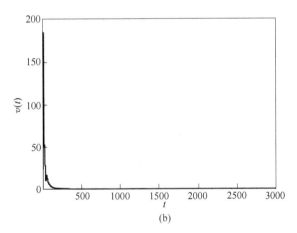

(b)

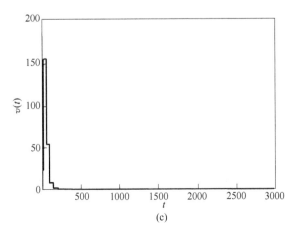

(c)

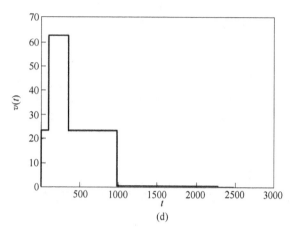

Fig. 5.4　Variance of $x(t)$ with different $\beta$
(a) $\beta=0.2$; (b) $\beta=1$; (c) $\beta=2$; (d) $\beta=4$

From Fig. 5.3, we can see all the states converge to the average value of the initial states, which is consistent with Theorem 5.4. From Fig. 5.2, we can see that $z_i(k)$, $i=1, \cdots, n$ converge faster as $\beta$ becomes larger. But $x_i(t)$, $i=1, \cdots, n$ doesn't always converge faster. The state $x(t)$ converges faster if $\beta$ changes from 0.2 to 1, which is shown in Fig. 5.3(a)(b). But the state $x(t)$ in Fig. 5.3(d) where $\beta$ becomes larger converges slower than that in Fig. 5.3(c). In other words, the convergence rate becomes faster and then slower with the increase of $\beta$, which can be seen from Fig. 5.4, too. The results are consistent with Theorem 5.5.

## 5.5　Notes

This chapter studies the average consensus problem of multi-agent systems with binary-valued communications under directed topologies. A two-time-scale algorithm alternating estimation and control has been constructed: estimate at the small-time scale and then design control at large-time scale. The estimation has been proved to be convergent and the convergence rate is given as the reciprocal of the holding time. The multi-agent system has been proved to asymptotically achieve average consensus and the consensus speed has been given.

There are many meaningful future works based on the results developed in this chapter. For example, can we omit the constraint on $\gamma$, which satisfies $0 < \gamma < \beta/2$ in the algorithm? How to design a best holding time $L_k$ to improve the convergence rate?

# 6 Consensus of Linear Multi-Agent Systems with Binary-Valued Measurements

The fourth and fifth chapters consider the consensus of first-order integrator systems. However, we encounter general linear systems in practical applications. The cooperative control of continuous-time linear multi-agent systems were studied in [51, 52]. Ref. [52] gave a necessary and sufficient condition for the consensusability of continuous-time linear multi-agent systems. For discrete-time linear systems, Ref. [53] proposed an observer-based distributed control protocol and gave a necessary and sufficient condition for consensusability under this control protocol. These works on consensus of linear multi-agent systems all require precise communication. In the case of quantized communication, Ref. [54] and Ref. [55] studied the consensusability for multi-agent systems of discrete-time linear dynamics with fixed topologies and switching topologies, respectively. Ref. [55] gave a class of control protocols based on encoding-decoding communication schemes under conditions that the system is observable and the system matrix is orthogonal. However, the consensusability of multi-agent systems with quantized communicational[54,55] require the quantization level to be greater than two. The binary-valued quantizers do not satisfy the requirement, since the quantization level is two.

In this chapter, we investigate the consensus problems of discrete-time linear multi-agent systems with binary-valued observations. The challenge of this chapter is to design state estimation dynamically by using binary-valued information. One natural idea is to use the parameter estimation methods for binary-valued observations in Section 4.2, which requires the states keep constant for a period of time. However, the states in linear multi-agent systems are dynamical. The recursive projection identification algorithm may provide another way to estimate the state. It is an online parameter estimation algorithm which does not require long time keeping the states unchanged. But it works well in Section 4.3 because the estimated state is almost unchanged when the system achieves consensus, which linear multi-agent systems do not satisfy.

In this chapter, we construct a two-step algorithm which can estimate the states of neighbors dynamically by using binary-valued information: At step 1, let the estimates have the same dynamics as the agents' states; At step 2, update the estimates by the recursive projection algorithm based on binary-valued information. Then, a stochastic approximation based consensus control is designed based on the estimates. In the closed-loop system, estimation and control are strong coupled. We analyze state estimation and consensus control together and finally give the convergence and convergence rate of the estimation and the control.

The rest of this chapter is organized as follows. In Section 6.1, we give the model of linear multi-agent systems and describe the consensus problem with binary-valued observations. In Section 6.2, we review some results on consensus of linear multi-agent systems with precise communication. In Section 6.3, we consider the case of binary-valued communication, including the consensus control algorithm and the main results of this chapter. In Section 6.4, simulations are given to demonstrate the theoretical results. In Section 6.5, we draw the concluding remarks and discuss some related future topics.

## 6.1 Problem Formulation

Consider a discrete-time linear multi-agent systems with $N$ agents as follows

$$x_i(t + 1) = Ax_i(t) + Bu_i(t), \quad i = 1, \cdots, N \quad (6.1)$$

where $A \in R^{n \times n}$ and $B \in R^{n \times m}$ are constant matrices, $x_i(t) \in R^n$ and $u_i(t) \in R^m$ are the state and the control of agent $i$ at time $t$, respectively.

Putting all the states in a vector, we can get $X(t) = [x_1^T(t), \cdots, x_N^T(t)]^T$. By Eq. (6.1), the multi-agent system can be written as

$$X(t + 1) = (I_N \otimes A)X(t) + (I_N \otimes B)u(t) \quad (6.2)$$

where $u(t) = [u_1^T(t), \cdots, u_N^T(t)]^T$. Let $\bar{X}(t) = \frac{1}{N}(\mathbb{1}_N \mathbb{1}_N^T \otimes I_n)X(t)$ and $\delta(t) = (\delta_1^T(t), \cdots, \delta_N^T(t))^T = X(t) - \bar{X}(t) = X(t) - \frac{1}{N}(\mathbb{1}_N \mathbb{1}_N^T \otimes I_n)X(t)$.

The agents are distributed by an undirected topology $G$. Each agent $i$ can only get binary-valued observations $s_{ij}(t)$ from its neighbors $j \in N_i$:

$$\begin{cases} y_{ij}(t) = x_j(t) + d_{ij}(t) \\ s_{ij}(t) = I_{\{y_{ij}(t) \leq c_{ij}\}} \end{cases} \quad (6.3)$$

where $x_j(t) \in R^n$ is the state of agent $j$, $d_{ij}(t) \in R^n$ is the noise, $y_{ij}(t) \in R^n$ is the unmeasurable output, $c_{ij} \in R^n$ is a given threshold vector, and $I_{\{a \leq c\}}$ is defined as

$$I_{|a\leqslant c|} = \left[I_{|a(k)\leqslant c(k)|} = \begin{cases} 1, & \text{if } a(k) \leqslant c(k) \\ 0, & \text{if } a(k) > c(k) \end{cases}\right]_{n\times 1}$$

for any $n$-dimensional vectors $a = [a(1), \cdots, a(n)]^T$ and $c = [c(1), \cdots, c(n)]^T$.

The goal of this chapter is to design control $u_i(t)$ based on the binary-valued observations $s_{ij}(t)$ to achieve consensus.

**Definition 6.1** The multi-agent system is defined to achieve consensus if there exists a control protocol such that

$$\lim_{t\to\infty} E\|x_i(t) - x_j(t)\| = 0, \quad i, j = 1, \cdots, N$$

or

$$\lim_{t\to\infty} E\|\delta(t)\| = 0$$

**Assumption 6.1** The system topology $G$ is connected.

**Assumption 6.2** The system matrix $A$ is orthogonal and $B$ satisfies $BB^T = I_n$.

**Remark 6.1** If the system matrix $A$ is neutrally stable, there exists a non-singular matrix $O$ such that $\tilde{A} = O^{-1}AO$ is a orthogonal matrix [56].

For the binary-valued communication (6.3), we have the following assumptions.

**Assumption 6.3** The noises $\{d_{ij}(t), (i, j) \in G, t = 1, 2, \cdots\}$ in system (6.3) are independent with respect to $i$, $j$, $t$ and identically distributed random variables with mean 0. The marginal distribution function and the associated marginal density function are respectively $[F_1(\cdot), \cdots, F_n(\cdot)]^T$ and $[f_1(\cdot), \cdots, f_n(\cdot)]^T$, where $f_i(x) = \dfrac{dF(x)}{dx} \neq 0$, for $i = 1, 2, \cdots, n$.

**Remark 6.2** In Assumption 6.3, we assume the marginal distribution function is known, which is an important issue in our approach to derive estimates of $x_i(t)$. If the distribution function is unknown, it can be estimated, see Ref. [57].

## 6.2 Review on the Case of Precise Communication

In this section, we review on the consensus problem of linear multi-agent systems with precise communication. In this situation, the states of the neighbors can be obtained accurately by the agent.

The control protocol is designed as

$$u_i(t) = K\sum_{j=1}^N p_{ij}(x_j(t) - x_i(t))$$

where $K \in R^{m\times n}$ is a constant matrix, $p_{ij}$ is the $i$th row $j$th column element of the adjacent matrix. It can be written as

$$u_i(t) = K \sum_{j \in N_i} (x_j(t) - x_i(t))$$

The control protocol can be written in a vector form:

$$u(t) = -(I_N \otimes K)(L \otimes I_n)X(t) \qquad (6.4)$$

where $L$ is the Laplacian matrix of $G$, i.e. $L(G)$.

By the control, the states of agents will be updated by

$$\begin{aligned}X(t+1) &= (I_N \otimes A - (I_N \otimes B)(I_N \otimes K)(L \otimes I_n))X(t) \\ &= (I_N \otimes A - L \otimes BK)X(t)\end{aligned}$$

Then, we can get

$$\begin{aligned}\delta(t+1) &= \left(I_N \otimes I_n - \frac{1}{N}\mathbb{1}_N \mathbb{1}_N^T \otimes I_n\right)X(t+1) \\ &\quad \left(I_N \otimes I_n - \frac{1}{N}\mathbb{1}_N \mathbb{1}_N^T \otimes I_n\right)(I_N \otimes A - L \otimes BK)X(t) \\ &\quad \left(I_N \otimes A - \frac{1}{N}\mathbb{1}_N \mathbb{1}_N^T \otimes A - L \otimes BK\right)X(t) \\ &\quad (I_N \otimes A - L \otimes BK)\left(I_N \otimes I_n - \frac{1}{N}\mathbb{1}_N \mathbb{1}_N^T \otimes I_n\right)X(t) \\ &\quad (I_N \otimes A - L \otimes BK)\delta(t) \qquad (6.5)\end{aligned}$$

**Proposition 6.1** [[54], Theorem 3.1]  Under Assumption 6.1, the multi-agent system (6.2) with control (6.4) can achieve consensus if and only if there exists a common control gain $K$ such that $\rho(A - \lambda_j BK) < 1$ for $j \in \{2, \cdots, N\}$, where $\lambda_2, \cdots, \lambda_N$ are the positive eigenvalues of Laplacian Matrix $L$.

By choosing $K = \mu B^T A$, Meng et al. [55] obtain the following proposition with Assumption 6.3.

**Proposition 6.2** [[55], Theorem 3.1]  Under Assumption 6.1 and 6.3, the multi-agent system (6.2) with control (6.4) can achieve consensus if

$$0 < \mu < \frac{1}{\lambda_N \lambda_{max}(A^T BB^T A)}$$

## 6.3 Case of Binary-Valued Communication

In this section, we consider the case of binary-valued communication, which means the communication between neighbor agents is binary-valued. Each agent cannot get the precise states of its neighbors, and what can be measured is binary-valued information given by (6.3). The consensus control is designed based on the binary-valued information.

## 6.3.1 Control Algorithm

In light of the case of precise communication, the control law is designed as follows:

$$u_i(t) = K_t \sum_{j \in N_i} (\hat{x}_{ij}(t) - x_i(t)), \quad i = 1, \cdots, N \tag{6.6}$$

where $K_t = \frac{1}{t} B^T A$, and $\hat{x}_{ij}(t)$ is the estimate of agent $j$'s state by agent $i$ at time $t$.

In the estimation, the recursive projection algorithm in Section 3.3 is proposed to deal with parameter estimation, where what to be estimate is a constant. The algorithm can be applied to the multi-agent systems in Section 4.3, where the state is static without control. However, the state has its own dynamics without control in the linear multi-agent system (6.1). The recursive projection algorithm cannot be easily applied to the linear multi-agent systems. In this chapter, the estimates are designed to be updated by two steps: First, the estimate is updated by the same dynamics as the true state without control; Second, the estimate is modified by the recursive projection algorithm using the binary-valued observations. Finally, the two-step estimation algorithm is given as follows:

$$\begin{cases} \tilde{x}_{ij}(t) = A\hat{x}_{ij}(t-1) \\ \hat{x}_{ij}(t) = \Pi_\Omega \left\{ \tilde{x}_{ij}(t) + \frac{\beta}{t}(\mathcal{F}(c_{ij} - \tilde{x}_{ij}(t)) - s_{ij}(t)) \right\} \end{cases}$$

which can be written as

$$\hat{x}_{ij}(t) = \Pi_\Omega \left\{ A\hat{x}_{ij}(t-1) + \frac{\beta}{t}(\mathcal{F}(c_{ij} - A\hat{x}_{ij}(t-1)) - s_{ij}(t)) \right\} \tag{6.7}$$

where $\beta$ is the step size for estimation updating, $\mathcal{F}(z) = [F_1(z_1), \cdots, F_n(z_n)]^T$ for any $z = [z_1, z_2, \cdots, z_n]^T$, $F_1(\cdot), F_2(\cdot), \cdots, F_n(\cdot)$ are the marginal distributed functions, $\Omega$ is a bounded set $\Omega = \{\omega \in R^n : |\omega| \leq M\}$ and $\Pi_\Omega(\cdot)$ is a projection mapping from $R^n$ to $\Omega$ defined as

$$\Pi_\Omega(\zeta) = \underset{\xi \in \Omega}{\arg\min} |\zeta - \xi|, \quad \forall \zeta \in R^n \tag{6.8}$$

**Remark 6.3** Due to the definition of the projection operator, we have

$$|\hat{x}_{ij}(t)| \leq M, \ j \in N_i, \ i = 1, \cdots, n$$

**Remark 6.4** [[17], Proposition 6] The projection function (6.8) satisfies

$$\| \Pi_\Omega(x_1) - \Pi_\Omega(x_2) \| \leq \| x_1 - x_2 \|, \text{ for any } x_1, x_2 \in R^n$$

In a word, the consensus protocol is as follows.

(1) Initiation: Let the initial state and the initial estimate of agent $i$ be respectively as follows:

$$x_i(1) = x_i^0, \quad \hat{x}_{ij}(0) = \hat{x}_{ij}^0$$

for $j \in N_i$, $i = 1, \cdots, N$, where $\| x_i^0 \| \leq M$, $\| \hat{x}_{ij}^0 \| \leq M$ and $M > 0$ is a given constant.

(2) Observation: Each agent $i$ gets the binary-valued observations from its neighbors:

$$\begin{cases} y_{ij}(t) = x_j(t) + d_{ij}(t) \\ s_{ij}(t) = I_{\{y_{ij}(t) \leq c_{ij}\}} \end{cases}$$

which is the same as Eq. (6.3).

(3) Estimation: Each agent $i$ estimates its neighbors' states by the two-step estimation algorithm:

$$\hat{x}_{ij}(t) = \Pi_\Omega \left\{ A\hat{x}_{ij}(t-1) + \frac{\beta}{t}(\mathcal{F}(c_{ij} - A\hat{x}_{ij}(t-1)) - s_{ij}(t)) \right\} \quad (6.9)$$

where $\beta$ is the step size for estimation updating, $\mathcal{F}(\cdot)$ is the vector of marginal distributed function with $\mathcal{F}(z) = [F_1(z_1), \cdots, F_n(z_n)]^T$, and $\Pi_\Omega(\cdot)$ is a projection operator defined in Eq. (6.8).

(4) Control: Based on the estimates, each agent designs the control by

$$u_i(t) = K_t \sum_{j \in N_i} (\hat{x}_{ij}(t) - x_i(t)), \quad i = 1, \cdots, N$$

where $K_t = \frac{1}{t} B^T A$.

(5) Repeat: Let $t = t+1$, go back to Step (2).

The consensus protocol can ensure the boundedness of the states.

**Proposition 6.3** The state of each agent is bounded by

$$\| x_i(t) \| \leq M, \quad i = 1, \cdots, N$$

if $t > N$.

**Proof** By Eq. (6.1) and Eq. (6.6) we have

$$x_i(t+1) = x_i(t) + \frac{1}{t} BB^T A \sum_{j \in N_i} (\hat{x}_{ij}(t) - x_i(t))$$

$$= x_i(t) + \frac{1}{t} A \sum_{j \in N_i} \hat{x}_{ij}(t) - \frac{1}{t} \sum_{j \in N_i} A x_i(t)$$

## 6.3 Case of Binary-Valued Communication

$$= 1 - \frac{|N_i|}{t} A x_i(t) + \frac{1}{t} \sum_{j \in N_i} A \hat{x}_{ij}(t)$$

where $|N_i|$ is the number of the elements in set $N_i$.

If $t = |N_i|$, then

$$x_i(|N_i| + 1) = \frac{1}{N_i} \sum_{j \in N_i} A \hat{x}_{ij}(t)$$

which implies

$$\| x_i(|N_i| + 1) \| \leq \frac{1}{|N_i|} \sum_{j \in N_i} \| A \| \, \| \hat{x}_{ij}(t) \| \leq M$$

Assume $\| x_i(p) \| \leq M$ for $p \geq |N_i| + 1$, then

$$\| x_i(p+1) \| \leq 1 - \frac{|N_i|}{t} \| A \| \, \| x_i(p) \| + \frac{1}{t} \sum_{j \in N_i} \| A \| \, \| \hat{x}_{ij}(t) \|$$

$$\leq 1 - \frac{|N_i|}{t} M + \frac{|N_i|}{t} M = M$$

By mathematical induce, we can get that

$$\| x_i(t) \| \leq M, \text{ if } t > |N_i|$$

If $t > N \geq \max_{i=1,\cdots,N} \{|N_i|\}$, then we have

$$\| x_i(t) \| \leq M, \ i = 1, \cdots, N$$

### 6.3.2 Main Results

By Eq. (6.1) and Eq. (6.6), the state is updated by

$$x_i(t+1) = x_i(t) + BK_t \sum_{j \in N_i} (\hat{x}_{ij}(t) - x_i(t))$$

$$= x_i(t) + BK_t \sum_{j \in N_i} (\hat{x}_{ij}(t) - x_j(t) + x_j(t) - x_i(t)) \quad (6.10)$$

which can be written in a vector form

$$X(t+1) = (I_N \otimes A) X(t) + (I_N \otimes BK_t) \epsilon(t) - (I_N \otimes BK_t)(L \otimes I_n) X(t)$$

$$= (I_N \otimes A - L \otimes BK_t) X(t) + (I_N \otimes BK_t) \epsilon(t)$$

$$(6.11)$$

where $\epsilon(t) = (\epsilon_1^T(t), \epsilon_2^T(t), \cdots, \epsilon_N^T(t))^T$, and $\epsilon_i(t) = \sum_{j \in N_i} (\hat{x}_{ij}(t) - x_j(t))$. Since

$$\delta(t) = \left( I_N \otimes I_n - \frac{1}{N} \mathbb{1}_N \mathbb{1}_N^T \otimes I_n \right) X(t)$$

It follows that

$$\delta(t+1) = \left(I_N \otimes I_n - \frac{1}{N} \mathbb{1}_N \mathbb{1}_N^T \otimes I_n\right) X(t+1)$$

$$= \left(I_N \otimes I_n - \frac{1}{N} \mathbb{1}_N \mathbb{1}_N^T \otimes I_n\right) ((I_N \otimes A - L \otimes BK_t) X(t) + (I_N \otimes BK_t) \epsilon(t))$$

$$= (I_N \otimes A - L \otimes BK_t) \delta(t) + \left(I_N \otimes I_n - \frac{1}{N} \mathbb{1}_N \mathbb{1}_N^T \otimes I_n\right) (I_N \otimes BK_t) \epsilon(t)$$

Let $\tilde{\delta}(t) = (U^T \otimes I_n) \delta(t)$, where $U$ is the orthogonal matrix $U = (e_1, e_2, \cdots, e_N)$ such that

$$U^T L U = \text{diag}(\lambda_1, \lambda_2, \cdots, \lambda_{N-1}, \lambda_N)$$

where $\lambda_1 = 0 < \lambda_2 < \lambda_3 < \cdots < \lambda_N$ are the eigenvalues of Laplacian Matrix $L$. Thus

$$\tilde{\delta}(t+1) = (U^T \otimes I_n) \delta(t+1)$$

$$= (U^T \otimes I_n) \Big[ (I_N \otimes A - L \otimes BK_t) \delta(t) +$$

$$(I_N \otimes I_n - \frac{1}{N} \mathbb{1}_N \mathbb{1}_N^T \otimes I_n)(I_N \otimes BK_t) \epsilon(t) \Big]$$

$$= (U^T \otimes I_n)(I_N \otimes A - L \otimes BK_t)(U^T \otimes I_n)^{-1} \tilde{\delta}(t) +$$

$$(U^T \otimes I_n)\left(I_N \otimes I_n - \frac{1}{N} \mathbb{1}_N \mathbb{1}_N^T \otimes I_n\right)(I_N \otimes BK_t) \epsilon(t)$$

$$= \text{diag}(A, A - \lambda_2 BK_t, \cdots, A - \lambda_N BK_t) \tilde{\delta}(t) +$$

$$\left(U^T \otimes BK_t - \frac{1}{N} U^T \mathbb{1}_N \mathbb{1}_N^T \otimes BK_t\right) \epsilon(t) \tag{6.12}$$

Let $K_t = \frac{1}{t} B^T A$, we have the following lemmas.

**Lemma 6.1** The consensus error satisfies

$$E(\delta^T(t)\delta(t)) \leq \left(1 - \frac{\lambda_2 - \nu}{t}\right) E(\delta^T(t-1)\delta(t-1)) + \frac{|D|^2}{t\nu} E(\epsilon^T(t)\epsilon(t)) + O\left(\frac{1}{t^2}\right)$$

where $\lambda_2$ is the smallest positive eigenvalue of Laplacian matrix $L$, $\nu$ is a positive constant, $D$ is defined as

$$D = \left((e_2^T, e_3^T, \cdots, e_N^T)^T - \frac{1}{N}(e_2^T, e_3^T, \cdots, e_N^T)^T \mathbb{1}_N \mathbb{1}_N^T\right) \otimes A$$

where $[e_1, e_2, \cdots, e_N] = U$ is the orthogonal matrix such that
$$U^T L U = \mathrm{diag}(\lambda_1, \lambda_2, \cdots, \lambda_{N-1}, \lambda_N)$$

**Proof** Since $\tilde{\delta}(t) = (U^T \otimes I_n)\delta(t)$, then
$$E(\delta^T(t)\delta(t)) = E(\tilde{\delta}^T(t)\tilde{\delta}(t))$$
Let $\tilde{\delta}(t) = [\tilde{\delta}_1^T(t) \in R^n, \tilde{\delta}_2^T(t) \in R^{n(N-1)}]^T$.
Then
$$\tilde{\delta}_1(t) = (e_1^T \otimes I_n)\delta(t) = \sum_{i=1}^{N} \delta_i(t) = 0$$
where $e_1 = (1, 1, \cdots, 1)^T \in R^n$. By (6.12), we have
$$\tilde{\delta}_2(t+1) = \mathrm{diag}(A - \lambda_2 B K_t, \cdots, A - \lambda_N B K_t)\tilde{\delta}_2(t) + D_t \epsilon(t)$$
$$= \mathrm{diag}\left(1 - \frac{\lambda_2}{t}, \cdots, 1 - \frac{\lambda_N}{t}\right) A \tilde{\delta}_2(t) + D_t \epsilon(t)$$
where the matrix $D_t$ is defined as
$$D_t = (e_2^T, e_3^T, \cdots, e_N^T)^T \otimes B K_t - \frac{1}{N}(e_2^T, e_3^T, \cdots, e_N^T)^T \mathbb{1}_N \mathbb{1}_N^T \otimes B K_t$$
$$= \frac{1}{t}\left((e_2^T, e_3^T, \cdots, e_N^T) - \frac{1}{N}(e_2^T, e_3^T, \cdots, e_N^T)^T \mathbb{1}_N \mathbb{1}_N^T\right) \otimes A = \frac{D}{t}$$
Then
$$\tilde{\delta}_2^T(t+1)\tilde{\delta}_2(t+1) = \tilde{\delta}_2^T(t) A^T \mathrm{diag}\left(1 - \frac{\lambda_2}{t}, \cdots, 1 - \frac{\lambda_N}{t}\right)^2 A \tilde{\delta}_2(t)$$
$$2\tilde{\delta}_2^T(t) A^T \mathrm{diag}\left(1 - \frac{\lambda_2}{t}, \cdots, 1 - \frac{\lambda_N}{t}\right) D_t \epsilon(t) + \epsilon^T(t) D_t^T D_t \epsilon(t)$$
$$= \tilde{\delta}_2^T(t) A^T \mathrm{diag}\left(\left(1 - \frac{\lambda_2}{t}\right)^2, \cdots, \left(1 - \frac{\lambda_N}{t}\right)^2\right) A \tilde{\delta}_2(t)$$
$$\frac{2}{t}\tilde{\delta}_2^T(t) A^T \mathrm{diag}\left(1 - \frac{\lambda_2}{t}, \cdots, 1 - \frac{\lambda_N}{t}\right) D \epsilon(t) + O\left(\frac{1}{t^2}\right)$$

When $t$ is large enough, we have
$$\mathrm{diag}\left(\left(1 - \frac{\lambda_2}{t}\right)^2, \cdots, \left(1 - \frac{\lambda_N}{t}\right)^2\right) \leq \left(1 - \frac{\lambda_2}{t}\right)^2 I_n$$

Hence
$$\tilde{\delta}_2^T(t+1)\tilde{\delta}_2(t+1) \leq \left(1 - \frac{\lambda_2}{t}\right)^2 \tilde{\delta}_2^T(t)\tilde{\delta}_2(t) +$$

$$\frac{2}{t}\tilde{\delta}_2^T(t)A^T\text{diag}\left(1-\frac{\lambda_2}{t},\cdots,1-\frac{\lambda_N}{t}\right)D\epsilon(t)+O\left(\frac{1}{t^2}\right) \qquad (6.13)$$

By Schwaz inequality, we have

$$\left(\tilde{\delta}_2^T(t)A^T\text{diag}\left(1-\frac{\lambda_2}{t},\cdots,1-\frac{\lambda_N}{t}\right)D\epsilon(t)\right)$$

$$\leq E(\tilde{\delta}_2^T(t)A^TA\tilde{\delta}_2(t))E\left(\epsilon^T(t)D^T\text{diag}\left(1-\frac{\lambda_2}{t},\cdots,1-\frac{\lambda_N}{t}\right)^2 D\epsilon^T(t)\right)$$

$$\leq E(\tilde{\delta}_2^T(t)\tilde{\delta}_2(t))\lambda_{\max}(D^TD)E(\epsilon^T(t)\epsilon(t))$$

$$= \nu E(\tilde{\delta}_2^T(t)\tilde{\delta}_2(t))\lambda_{\max}(D^TD)E(\epsilon^T(t)\epsilon(t))/\nu$$

$$\leq \frac{1}{2}(\nu E(\tilde{\delta}_2^T(t)\tilde{\delta}_2(t))+\lambda_{\max}(D^TD)E(\epsilon^T(t)\epsilon(t))/\nu)$$

Taking expectation on both side of Eq. (6.13) we can get

$$(\tilde{\delta}_2^T(t+1)\tilde{\delta}_2(t+1))$$

$$\leq \left(1-\frac{\lambda_2}{t}\right)^2 E(\tilde{\delta}_2^T(t)\tilde{\delta}_2(t))+\frac{\nu}{t}E(\tilde{\delta}_2^T(t)\tilde{\delta}_2(t))+\frac{\lambda_{\max}(D^TD)}{ta_1}E(\epsilon^T(t)\epsilon(t))+O\left(\frac{1}{t^2}\right)$$

$$= \left(1-\frac{\lambda_2-\nu}{t}\right)E(\tilde{\delta}_2^T(t)\tilde{\delta}_2(t))+\frac{|D|^2}{t\nu}E(\epsilon^T(t)\epsilon(t))+O\left(\frac{1}{t^2}\right)$$

Since

$$E(\delta^T(t)\delta(t))=E(\tilde{\delta}^T(t)\tilde{\delta}(t))=E(\tilde{\delta}_1^T(t)\tilde{\delta}_1(t))+E(\tilde{\delta}_2^T(t)\tilde{\delta}_2(t))=E(\tilde{\delta}_2^T(t)\tilde{\delta}_2(t))$$

the lemma can be proved.

Let $\varepsilon_{ij}(t)=\hat{x}_{ij}(t)-x_j(t)$. Putting $\varepsilon_{ij}(t)$, $j\in N_i$, $i=1,\cdots,N$ in a given order yields the error vector $\varepsilon(t)$. Without loss of generality, let

$$\varepsilon(t)=(\varepsilon_{1r_1}^T(t),\cdots,\varepsilon_{1r_{d_1}}^T(t),\varepsilon_{2r_{d_1+1}}^T(t),\cdots,\varepsilon_{2r_{d_1+d_2}}^T(t),\cdots,$$

$$\varepsilon_{Nr_{d_1+\cdots+d_{N-1}+1}}^T(t),\cdots,\varepsilon_{Nr_{d_1+\cdots+d_N}}^T(t))^T \qquad (6.14)$$

where $r_1,\cdots,r_{d_1}\in N_1$, $r_{d_1+1},\cdots,r_{d_1+d_2}\in N_2,\cdots,r_{d_1+\cdots+d_{N-1}+1},\cdots,r_{d_1+\cdots+d_N}\in N_N$, $d_i$ is the degree of node $i$. Assume $d_1+\cdots+d_N=h$, then $\varepsilon(t)$ is $nh$-dimensional.

**Lemma 6.2** The mean square error of the estimation satisfies

$$E(\varepsilon^T(t)\varepsilon(t))\leq\left(1-\frac{2\beta f_M-2\|Q\|\|P\|-\|Q\|^2\lambda_N^2/\gamma}{t}\right)E(\varepsilon^T(t-1)\varepsilon(t-1))+$$

$$\frac{\gamma}{t}E(\delta^T(t-1)\delta(t-1))+O\left(\frac{1}{t^2}\right)$$

where $f_M = \min_{k=1,\cdots,n} f_k(\max_{j \in N_i, i=1,\cdots,N} c_{ij}(k) + M)$, $c_{ij} = [c_{ij}(1), \cdots, c_{ij}(n)]^T$ is the threshold vector, $f_1(\cdot), \cdots, f_n(\cdot)$ are the marginal density functions of the noise, $L$ is the Laplacian matrix, $\gamma$ is a positive constant, matrices $P$ and $Q$ are defined as follows:

$$P = \begin{bmatrix} \underbrace{11\cdots1}_{d_1} & & & \\ & \underbrace{11\cdots1}_{d_2} & & 0 \\ & 0 & \cdots & \\ & & & \underbrace{11\cdots1}_{d_N} \end{bmatrix}_{N \times h} \quad \text{and} \quad Q = \begin{bmatrix} q_{1r_1}^T \\ \vdots \\ q_{1r_{d_1}}^T \\ q_{2r_{d_1+1}}^T \\ \vdots \\ q_{Nr_{d_1+\cdots+d_N}}^T \end{bmatrix}_{h \times N} \quad (6.15)$$

where $q_{ij}$ is a $N$-dimensional vector

$$q_{ij} = (0, \cdots, 0, \underset{j\text{th position}}{1}, 0, \cdots, 0)^T$$

**Proof** By Proposition 6.3, we have $x_i(t) \in \Omega$ for $i = 1, \cdots, N$. By estimation Eq. (6.7), state updating Eq. (6.10) and Remark 6.4, we can get that

$$\|\hat{x}_{ij}(t) - x_j(t)\|$$

$$\leq \left\| A\hat{x}_{ij}(t-1) + \frac{\beta}{t}(\mathcal{F}(c_{ij} - A\hat{x}_{ij}(t-1)) - s_{ij}(t)) - x_j(t) \right\|$$

$$= \left\| A\hat{x}_{ij}(t-1) + \frac{\beta}{t}(\hat{\mathcal{F}}_{ij}(t-1) - s_{ij}(t)) - Ax_j(t-1) - BK_t \sum_{p \in N_j}(\hat{x}_{jp}(t-1) - x_p(t-1)) + BK_t \sum_{p \in N_j}(x_j(t-1) - x_p(t-1)) \right\|$$

$$= \left\| A\varepsilon_{ij}(t-1) + \frac{\beta}{t}(\hat{\mathcal{F}}_{ij}(t-1) - s_{ij}(t)) - \frac{A}{t}\sum_{p \in N_j}\varepsilon_{jp}(t-1) + \frac{A}{t}\sum_{p \in N_j}(\delta_j(t-1) - \delta_p(t-1)) \right\| \quad (6.16)$$

where $\hat{\mathcal{F}}_{ij}(t-1) = \mathcal{F}(c_{ij} - A\hat{x}_{ij}(t-1))$. Let

$$\eta_{ij}(t-1) = A\varepsilon_{ij}(t-1) + \frac{\beta}{t}(\hat{\mathcal{F}}_{ij}(t-1) - s_{ij}(t)) - \frac{A}{t}\sum_{p \in N_j}\varepsilon_{jp}(t-1) + \frac{A}{t}\sum_{p \in N_j}(\delta_j(t-1) - \delta_p(t-1))$$

putting $\eta_{ij}(t-1)$, $j \in N_i$, $i = 1, \cdots, n$ in the same order as $\varepsilon(t)$ gives the vector $\eta(t-1)$, as follows

$$\eta(t-1) = (I_h \otimes A)\varepsilon(t-1) + \frac{\beta}{t}\Phi_{FS}(t) - \frac{1}{t}(I_h \otimes A)(Q \otimes I_n)(P \otimes I_n)\varepsilon(t-1) +$$

$$\frac{1}{t}(I_h \otimes A)(Q \otimes I_n)(L \otimes I_n)\delta(t-1)$$

$$= (I_h \otimes A)\varepsilon(t-1) + \frac{\beta}{t}\Phi_{FS}(t) - \frac{1}{t}(QP \otimes A)\varepsilon(t-1) +$$

$$\frac{1}{t}(QL \otimes A)\delta(t-1) \tag{6.17}$$

where $\Phi_{FS}(t)$ is the vector which putting $(\hat{\mathcal{F}}_{ij}(t-1) - s_{ij}(t))$ in the same order as $\varepsilon(t)$.

By Eq. (6.16) and Eq. (6.17), we have

$$E(\varepsilon^T(t)\varepsilon(t)) \leq E(\eta^T(t-1)\eta(t-1))$$

$$= E(\varepsilon^T(t-1)\varepsilon(t-1)) + \frac{2\beta}{t}E(\varepsilon^T(t-1)(I_h \otimes A)^T\Phi_{FS}(t)) -$$

$$\frac{2}{t}E(\varepsilon^T(t-1)(QP \otimes I_n)\varepsilon(t-1)) +$$

$$\frac{2}{t}E(\varepsilon^T(t-1)(QL \otimes I_n)\delta(t-1)) + o\left(\frac{1}{t^2}\right) \tag{6.18}$$

By Eq. (6.3), we have

$$E(s_{ij}(t)) = \mathcal{F}(c_{ij} - x_j(t))$$

Then,

$$E(\varepsilon_{ij}^T(t-1)A^T(\hat{\mathcal{F}}_{ij}(t-1) - s_{ij}(t)))$$

$$= E(\varepsilon_{ij}^T(t-1)A^T(\mathcal{F}(c_{ij} - A\hat{x}_{ij}(t-1)) - \mathcal{F}(c_{ij} - x_j(t))))$$

$$= -E(\varepsilon_{ij}^T(t-1)A^T)\Lambda(f(\xi_{ij}))(A\hat{x}_{ij}(t-1) - x_j(t))$$

where $\xi_{ij}$ is in the middle of $c_{ij} - A\hat{x}_{ij}(t-1)$ and $c_{ij} - x_j(t)$, and $\Lambda(f(\xi_{ij})) = \mathrm{diag}\{f_1(\xi_{ij}(1)), \cdots, f_n(\xi_{ij}(n))\}$. Under Assumption 6.2, Remark 6.3 and Proposition 6.3, we have

$$f_k(\xi_{ij}(k)) \geq f_k(\max_{j \in N_i, i=1, \cdots, N}\{c_{ij}(k)\} + M)$$

$$\geq \min_{k=1, \cdots, n} f_k(\max_{j \in N_i, i=1, \cdots, N}\{c_{ij}(k)\} + M) \triangleq f_M$$

Moreover

$$E(\varepsilon_{ij}^T(t-1)A^T(\hat{\mathcal{F}}_{ij}(t-1) - s_{ij}(t)))$$
$$\leq -f_M E(\varepsilon_{ij}^T(t-1)A^T(A\hat{x}_{ij}(t-1) - x_j(t)))$$
$$= -f_M E(\varepsilon_{ij}^T(t-1)A^T(A\hat{x}_{ij}(t-1) - Ax_j(t-1) - \frac{A}{t}\sum_{p \in N_j}(\hat{x}_{jp}(t-1) - x_p(t-1))))$$
$$= -f_M E(\varepsilon_{ij}^T(t-1)\varepsilon_{ij}(t-1)) + O\left(\frac{1}{t}\right)$$

which implies

$$\frac{2\beta}{t}E(\varepsilon^T(t-1)(I_h \otimes A)^T \Phi_{FS}(t)) \leq -\frac{2\beta f_M}{t}E(\varepsilon^T(t-1)\varepsilon(t-1)) + O\left(\frac{1}{t^2}\right) \quad (6.19)$$

Together with

$$-\frac{2}{t}E(\varepsilon^T(t-1)(QP \otimes I_n)\varepsilon(t-1))$$
$$\leq \frac{2}{t}\sqrt{E(\varepsilon^T(t-1)(Q^TQ \otimes I_n)\varepsilon(t-1))E(\varepsilon^T(t-1)(P^TP \otimes I_n)\varepsilon(t-1))}$$
$$\leq \frac{2}{t}\sqrt{\lambda_{max}(Q^TQ)\lambda_{max}(P^TP)}\, E(\varepsilon^T(t-1)\varepsilon(t-1))$$
$$= \frac{2}{t}\|Q\|\|P\|E(\varepsilon^T(t-1)\varepsilon(t-1))$$

and

$$\frac{2}{t}E(\varepsilon^T(t-1)(QL \otimes I_n)\delta(t-1))$$
$$\leq \frac{2}{t}\sqrt{E(\varepsilon^T(t-1)(L^TQ^TQL \otimes I_n)\varepsilon(t-1))E(\delta^T(t-1)\delta(t-1))}$$
$$\leq \frac{2}{t}\sqrt{\frac{1}{\gamma}\lambda_{max}(Q^TQ)\lambda_{max}(L^TL)E(\varepsilon^T(t-1)\varepsilon(t-1))\gamma E(\delta^T(t-1)\delta(t-1))}$$
$$\leq \frac{1}{t}\left(\frac{1}{\gamma}\|Q\|^2\|L\|^2 E(\varepsilon^T(t-1)\varepsilon(t-1)) + \gamma E(\delta^T(t-1)\delta(t-1))\right)$$

We can obtain

$$E(\varepsilon^T(t)\varepsilon(t)) \leq \left(1 - \frac{2\beta f_M - 2\|Q\|\|P\| - \|Q\|^2\|L\|^2/\gamma}{t}\right).$$

$$E(\varepsilon^T(t-1)\varepsilon(t-1)) + \frac{\gamma}{t}E(\delta^T(t-1)\delta(t-1)) + O\left(\frac{1}{t^2}\right)$$

**Theorem 6.1** (Convergence) Under Assumptions 6.1 ~ 6.3, the mean square error of estimation Eq. (6.9) will converge to zero and the states of system (6.1) will achieve consensus:

$$E(\varepsilon^T(t)\varepsilon(t)) \to 0 \quad \text{and} \quad E(\delta^T(t)\delta(t)) \to 0 \tag{6.20}$$

if the coefficient $\beta$ is selected by

$$\beta > \frac{1}{f_M}\left(\frac{4\|D\|^4\|P\|^4}{\lambda_2^3} + \frac{\lambda_2\lambda_N^2\|Q\|}{4\|D\|^2\|P\|^2} + \|Q\|\|P\|\right)$$

where $\beta, f_M, \lambda_2, \lambda_N$ are defined in Lemmas 6.1 and 6.2, $P$ and $Q$ are defined by Eq. (6.15).

**Proof** Let

$$R(t) = E(\varepsilon^T(t)\varepsilon(t)), \quad V(t) = E(\delta^T(t)\delta(t))$$

By Lemma 6.2, we have

$$R(t) \leq \left(1 - \frac{2\beta f_M - 2\|Q\|\|P\| - \|Q\|^2\lambda_N^2/\gamma}{t}\right)R(t-1) + \frac{\gamma}{t}V(t-1) + O\left(\frac{1}{t^2}\right) \tag{6.21}$$

Since $\epsilon_i(t) = \sum_{j \in N_i}(\hat{x}_{ij}(t) - x_j(t))$, it follows that $\epsilon(t) = (P \otimes I_n)\varepsilon(t)$. By Lemma 6.1, we have

$$V(t) \leq \left(1 - \frac{\lambda_2 - \nu}{t}\right)V(t-1) + \frac{\|D\|^2\|P\|^2}{t\nu}R(t-1) + O\left(\frac{1}{t^2}\right)$$

Let $\nu = \frac{\lambda_2}{2}$, then we have

$$V(t) \leq \left(1 - \frac{\lambda_2}{2t}\right)V(t-1) + \frac{2\|D\|^2\|P\|^2}{t\lambda_2}R(t-1) + O\left(\frac{1}{t^2}\right) \tag{6.22}$$

Let $Z(t) = (R(t), V(t))^T$, By Eq. (6.21) and Eq. (6.22) we can get that

$$|Z(t)| \leq \left\|\left(I - \frac{W}{t}\right)Z(t-1) + O\left(\frac{1}{t^2}\right)\right\| \tag{6.23}$$

where

$$W = \begin{pmatrix} 2\beta f_M - 2\|Q\|\|P\| - \|Q\|^2\lambda_N^2/\gamma & -\gamma \\ -\frac{2\|D\|^2\|P\|^2}{\lambda^2} & \frac{\lambda_2}{2} \end{pmatrix}$$

## 6.3 Case of Binary-Valued Communication

Let $\gamma = \dfrac{2\|D\|^2\|P\|^2}{\lambda_2}$, the matrix $W$ will be an a symmetric matrix, and

$$\lambda\left(I - \frac{2W}{t} + \frac{W^2}{t^2}\right) = \left(1 - \frac{\lambda(W)}{t}\right)^2 \leqslant \left(1 - \frac{\lambda_{min}(W)}{t}\right)^2, \text{ as } t > \lambda_{max}(W)$$

which implies

$$\left\|I - \frac{W}{t}\right\| = \sqrt{\lambda_{max}\left(I - \frac{2W}{t} + \frac{W^2}{t^2}\right)} \leqslant 1 - \frac{\lambda_{min}(W)}{t}$$

Consequently, we have by Eq. (6.23)

$$\|Z(t)\| \leqslant \left(1 - \frac{\lambda_{min}(W)}{t}\right)\|Z(t-1)\| + \frac{\|H\|}{t^2}$$

$$\leqslant \prod_{i=1}^{t}\left(1 - \frac{\lambda_{min}(W)}{i}\right)\|Z(0)\| + \sum_{i=1}^{t}\prod_{l=i+1}^{t}\left(1 - \frac{\lambda_{min}(W)}{l}\right)\frac{\|H\|}{i^2}$$

According to Theorem 2.10, it follows that

$$\|Z(t)\| = \begin{cases} O\left(\dfrac{1}{t^{\lambda_{min}(W)}}\right) & \lambda_{min}(W) < 1 \\ O\left(\dfrac{\ln t}{t}\right) & \lambda_{min}(W) = 1 \\ O\left(\dfrac{1}{t}\right) & \lambda_{min}(W) > 1 \end{cases} \quad (6.24)$$

which implies

$$\lim_{t \to \infty} \|Z(t)\| = 0, \quad \lambda_{min}(W) > 0 \quad (6.25)$$

Let

$$a = 2\beta f_M - 2\|Q\|\|P\| - \frac{\|Q\|^2\lambda_N^2\lambda_2}{2\|D\|^2\|P\|^2}, \quad b = -\frac{2\|D\|^2\|P\|^2}{\lambda_2}, \quad c = \frac{\lambda_2}{2}$$

then the matrix $W$ can be written as

$$W = \begin{pmatrix} a & b \\ b & c \end{pmatrix}$$

Letting $|\lambda I - W| = 0$ gives the eigenvalues of matrix $W$ as follows

$$\lambda_{max}(W) = \frac{a + c + \sqrt{(a+c)^2 - 4(ac - b^2)}}{2}$$

$$\lambda_{min}(W) = \frac{a + c - \sqrt{(a+c)^2 - 4(ac - b^2)}}{2} \quad (6.26)$$

If

$$\beta > \frac{1}{f_M}\left(\frac{4\|D\|^4\|P\|^4}{\lambda_2^3} + \frac{\lambda_2\lambda_N^2\|Q\|^2}{4\|D\|^2\|P\|^2} + \|Q\|\|P\|\right)$$

then $ac > b^2$. By Eq. (6.26), we have $\lambda_{\min}(W) > 0$. Hence, we can get the theorem by Eq. (6.25).

**Theorem 6.2** (Convergence rate)  Under Assumptions 6.1 ~ 6.3, we can get the convergence rate of estimation and consensus speed as follows

$$E(\varepsilon^T(t)\varepsilon(t)) = O\left(\frac{1}{t}\right), \quad E(\delta^T(t)\delta(t)) = O\left(\frac{1}{t}\right)$$

if the coefficient $\beta$ is selected by $\beta > \dfrac{\gamma_2}{2f_M}$ and $\lambda_2 > 2$, where $\beta$ and $f_M$ are defined in Lemmas 6.1 and 6.2.

**Proof**  Let $W = \begin{pmatrix} a & b \\ b & c \end{pmatrix}$, where $a = 2\beta f_M - 2\|Q\|\|P\| - \dfrac{\|Q\|^2\lambda_N^2\lambda_2}{2\|D\|^2\|P\|^2}$, $b = -\dfrac{2\|D\|^2\|P\|^2}{\lambda_2}$, $c = \dfrac{\lambda_2}{2}$. Then

$$\|Z(t)\| \leq \left(1 - \frac{\lambda_{\min}(W)}{t}\right)\|Z(t-1)\| + \frac{\|H\|}{t^2}$$

$$\leq \prod_{i=1}^{t}\left(1 - \frac{\lambda_{\min}(W)}{i}\right)\|Z(0)\| + \sum_{i=1}^{t}\prod_{l=i+1}^{t}\left(1 - \frac{\lambda_{\min}(W)}{l}\right)\frac{\|H\|}{i^2}$$

If

$$\beta > \frac{1}{f_M}\left(\frac{4\|D\|^4\|P\|^4}{\lambda_2^2(\lambda_2 - 2)} + \frac{\lambda_2\lambda_N^2\|Q\|^2}{4\|D\|^2\|P\|^2} + \|Q\|\|P\| + \frac{1}{2}\right)$$

and $\lambda_2 > 2$, then we have

$$a > 1, \quad c > \frac{b^2}{a-1} + 1$$

Consequently,

$$(a-1)c > b^2 + (a-1)$$

which implies

$$\sqrt{(a+c)^2 - 4(ac - b^2)} < a + c - 2$$

Hence, the smallest eigenvalue of matrix $W$ follows

$$\lambda_{\min}(W) = \frac{(a+c) - \sqrt{(a+c)^2 - 4(ac - b^2)}}{2} > \frac{(a+c) - (a+c-2)}{2} = 1$$

By Eq. (6.24), we can get

$$Z(t) = O\left(\frac{1}{t}\right)$$

which proves the theorem.

## 6.4  Numerical Simulation

We give a multi-agent system with five agents as an example to justify the previous theories. In the multi-agent system, the agents' states are two-dimensional vectors which are updated by

$$x_i(t+1) = Ax_i(t) + Bu_i(t), \quad i = 1, \cdots, 5 \tag{6.27}$$

where $x_i(t) \in R^2$, $u_i(t) \in R^2$, $A = \begin{bmatrix} 1 & 0 \\ 0 & 1 \end{bmatrix}$ and $B = \begin{bmatrix} 1 & 0 \\ 0 & 1 \end{bmatrix}$.

The five agents communicate with each other in a fixed link as Fig. 6.1. The corresponding Laplacian matrix is

$$L = \begin{bmatrix} 3 & -1 & 0 & -1 & -1 \\ -1 & 2 & -1 & 0 & 0 \\ 0 & -1 & 2 & -1 & 0 \\ -1 & 0 & -1 & 3 & -1 \\ -1 & 0 & 0 & -1 & 2 \end{bmatrix}$$

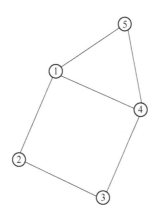

Fig. 6.1  The topology of five agents

And each agent can only obtain the binary-valued information from its neighbors

$$\begin{cases} y_{ij}(t) = x_j(t) + d_{ij}(t) \\ s_{ij}(t) = I_{\{y_{ij}(t) \leq 0\}} \end{cases}$$

where $d_{ij}(t) \in R^2$ is the noise generated by joint normal distribution $N(0, 0, 36, 36, 0)$, the threshold $c_{ij}$ equals 0, and $s_{ij}(t) \in R^2$ is the binary-valued observation.

In this simulation, we set the initial states as $x(1) = [(0, 5)^T, (5, 0)^T, (0, -5)^T, (-5, 0)^T, (0, 0)^T]^T$, and let the initial estimates as $\hat{x}(0) = [\hat{x}_{12}(0), \hat{x}_{14}(0), \hat{x}_{15}(0), \hat{x}_{21}(0), \hat{x}_{23}(0), \hat{x}_{32}(0), \hat{x}_{34}(0), \hat{x}_{41}(0), \hat{x}_{43}(0), \hat{x}_{45}(0), \hat{x}_{51}(0), \hat{x}_{54}(0)]^T = \mathbb{1}_{12} \otimes (1, 1)^T$, where $\mathbb{1}_{12}$ is a 12-dimensional vector with all the elements being one. Each agent estimates its neighbors' states by (6.7) with $M = 8$ and $\beta = 800$, then designs the control by (6.6) with $K_t = \frac{1}{t} B^T A$. Finally, the five agents update their states by (6.27). Fig. 6.2 and Fig. 6.3 show respectively the trajectories of the first and the second component of five agents' states, which demonstrate that the five agents can achieve consensus. Fig. 6.4 gives the linear tendency of log-varianve, implying the convergence speed of $O(1/t)$, which is consistent with Theorem 6.2.

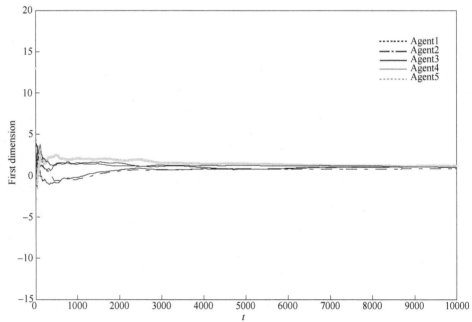

Fig. 6.2  The trajectories of the first component of five agents' states

## 6.5 Notes

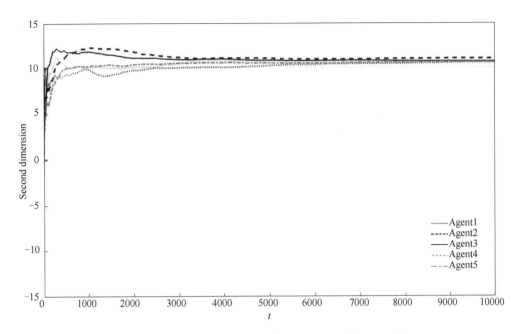

Fig. 6.3  The trajectories of the second component of five agents' states

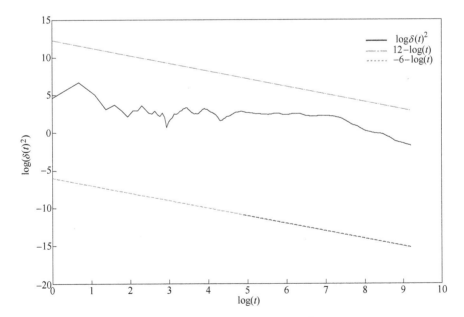

Fig. 6.4  The linear property of $\log(\delta^2(t))$

## 6.5 Notes

In this chapter, consensus control of discrete-time linear multi-agent systems with binary-valued communication has been considered. A two-step algorithm is designed for each agent to estimate its neighbors' states dynamically. A consensus control is proposed based on the estimates. It is proved that, for a connected network, by properly selecting the coefficient in the estimation algorithm, the estimates can converge to the true states and the states of all the agents can achieve consensus. Moreover, the convergence rate of the estimation and the consensus speed are both proved to be $O(1/t)$.

In this chapter, we assume that the system matrix is an orthogonal matrix. The consensus problem of linear multi-agent systems with more general system matrix is a more challenging topic for future research. Since link failures are inevitable in many practical communication networks, how to deal with the case of link failures is also interesting.

## References

[1] Zhang J F, Yin G G. System identification using binary sensors [J]. IEEE Transactions on Automatic Control, 2003, 48 (11): 1892~1907.

[2] Bi W, Kang G, Zhao Y, et al. SVSI: Fast and Powerful Set-Valued System Identification Approach to Identifying Rare Variants in Sequencing Studies for Ordered Categorical Traits [J]. Annals of human genetics, 2015, 79 (4): 294~309.

[3] Wang T, Bi W, Zhao Y, et al. Radar target recognition algorithm based on RCS observation sequence—set-valued identification method [J]. Journal of Systems Science and Complexity, 2016, 29 (3): 573~588.

[4] Olfati-Saber R. Distributed Kalman filter with embedded consensus filters [C]. In Proceedings of the 44th IEEE Conference on Decision and Control, 2005: 8179~8184.

[5] Ren W, Beard R W. Distributed consensus in multi-vehicle cooperative control [M]. London: Springer London, 2008.

[6] Bhatta P, Leonard N E. Stabilization and coordination of underwater gliders [C]. In Proceedings of the 41st IEEE Conference on Decision and Control, 2002, 2: 2081~2086.

[7] Desai J P, Kumar V, Ostrowski J P. Control of changes in formation for a team of mobile robots [C]. In Proceedings 1999 IEEE International Conference on Robotics and Automation (Cat. No. 99CH36288C). 1999, 2: 1556~1561.

[8] Jadbabaie A, Lin J, Morse A S. Coordination of groups of mobile autonomous agents using nearest neighbor rules [J]. IEEE Transactions on automatic control, 2003, 48 (6): 988~1001.

[9] Zhao Y, Wang L Y, Yin G G, et al. Identification of Wiener systems with binary-valued output observations [J]. Automatica, 2007, 43 (10): 1752~1765.

[10] Zhao Y, Zhang J F, Wang L Y, et al. Identification of Hammerstein systems with quantized observations [J]. SIAM Journal on Control and Optimization, 2010, 48 (7): 4352~4376.

[11] Wang L Y, Yin G G. Asymptotically efficient parameter estimation using quantized output observations [J]. Automatica, 2007, 43 (7): 1178~1191.

[12] Godoy B I, Goodwin G C, Agüero J C, et al. On identification of FIR systems having quantized output data [J]. Automatica, 2011, 47 (9): 1905~1915.

[13] Zhao Y, Bi W, Wang T. Iterative parameter estimate with batched binary-valued observations [J]. Science China Information Sciences, 2016, 59 (5): 1~18.

[14] Jafari K, Juillard J, Colinet E. A recursive system identification method based on binary measurements [C]. In Proceedings of 49th IEEE Conference on Decision and Control (CDC), 2010: 1154~1158.

[15] You K. Recursive algorithms for parameter estimation with adaptive quantizer [J]. Automatica, 2015, 52: 192~201.

[16] Bourgois L, Juillard J. Convergence analysis of an online approach to parameter estimation problems based on binary noisy observations [C]. In Proceedings of the 51st IEEE Conference on Decision and Control (CDC), 2012: 1506~1511.

[17] Guo J, Zhao Y. Recursive projection algorithm on FIR system identification with binary-valued observations [J]. Automatica, 2013, 49 (11): 3396~3401.

[18] Ren W, Beard R W, Kingston D B. Multi-agent Kalman consensus with relative uncertainty [C]. In Proceedings of the 2005, American Control Conference, 2005: 1865~1870.

[19] Huang M, Manton J H. Stochastic Lyapunov analysis for consensus algorithms with noisy measurements [C]. In Proceedings of the 2007 American Control Conference, 2007: 1419~1424.

[20] Kar S, Moura J M F. Distributed average consensus in sensor networks with random link failures and communication channel noise [C]. In Proceedings of the 2007 Conference Record of the Forty-First Asilomar Conference on Signals, Systems and Computers, 2007: 676~680.

[21] Li T, Zhang J F. Mean square average-consensus under measurement noises and fixed topologies: Necessary and sufficient conditions [J]. Automatica, 2009, 45 (8): 1929~1936.

[22] Kashyap A, Basar T, Srikant R. Consensus with quantized information updates [C]. In Proceedings of the 45th IEEE Conference on Decision and Control, 2006: 2728~2733.

[23] Kashyap A, Başar T, Srikant R. Quantized consensus [J]. Automatica, 2007, 43 (7): 1192~1203.

[24] Frasca P, Carli R, Fagnani F, et al. Average consensus on networks with quantized communication [J]. International Journal of Robust and Nonlinear Control: IFAC - Affiliated Journal, 2009, 19 (16): 1787~1816.

[25] Liu S, Li T, Xie L, et al. Continuous-time and sampled-data-based average consensus with logarithmic quantizers [J]. Automatica, 2013, 49 (11): 3329~3336.

[26] Carli R, Bullo F, Zampieri S. Quantized average consensus via dynamic coding/decoding schemes [J]. International Journal of Robust and Nonlinear Control: IFAC-Affiliated Journal, 2010, 20 (2): 156~175.

[27] Li T, Fu M, Xie L, et al. Distributed consensus with limited communication data rate [J]. IEEE Transactions on Automatic Control, 2010, 56 (2): 279~292.

[28] Li T, Xie L. Distributed coordination of multi-agent systems with quantized-observer based encoding-decoding [J]. IEEE Transactions on Automatic Control, 2012, 57 (12): 3023~3037.

[29] Yu H, Bi W, Liu C, et al. Protein-interaction-network-based analysis for genome-wide association analysis of schizophrenia in Han Chinese population [J]. Journal of Psychiatric Research, 2014, 50: 73~78.

[30] Chow Y S, Teicher H. Probability theory: independence, interchangeability, martingales [M]. Springer Science & Business Media, 2012.

[31] Fang B R, Zhou J D, Li Y M. Matrix Theory [M]. Tsinghua University Press, 2004 (in Chinese).

[32] Sheng S, Xie S Q, Pan C Y. Probability Theory and Mathematical Statistics, Fourth Edition [J]. Higher Education Press, 2008 (in Chinese).

[33] Mesbahi M, Egerstedt M. Graph theoretic methods in multiagent networks [M]. Princeton University Press, 2010.

[34] Aguero J C, Goodwin G C, Yuz J I. System identification using quantized data [C]. In Pro-

ceedings of the 46th IEEE Conference on Decision and Control, 2007: 4263~4268.

[35] Casini M, Garulli A, Vicino A. Time complexity and input design in worst-case identification using binary sensors [C]. In Proceedings of the 46th IEEE Conference on Decision and Control, 2007: 5528~5533.

[36] You K, Xie L, Sun S, et al. Multiple-level quantized innovation Kalman filter [J]. IFAC Proceedings Volumes, 2008, 41 (2): 1420~1425.

[37] Guo J, Wang L Y, Yin G G, et al. Asymptotically efficient identification of FIR systems with quantized observations and general quantized inputs [J]. Automatica, 2015, 57: 113~122.

[38] Guo J, Zhao Y. Recursive projection algorithm on FIR system identification with binary-valued observations [J]. Automatica, 2013, 49 (11): 3396~3401.

[39] Ren W, Cao Y. Distributed coordination of multi-agent networks: emergent problems, models, and issues [M]. Springer Science & Business Media, 2010.

[40] Cao J, Spooner D P, Jarvis S A, et al. Grid load balancing using intelligent agents [J]. Future generation computer systems, 2005, 21 (1): 135~149.

[41] Tanner H G, Jadbabaie A, Pappas G J. Flocking in fixed and switching networks [J]. IEEE Transactions on Automatic control, 2007, 52 (5): 863~868.

[42] Reynolds C W. Flocks, herds and schools: A distributed behavioral model [C]. In Proceedings of the 14th annual conference on Computer graphics and interactive techniques, 1987: 25~34.

[43] Tanner H G, Jadbabaie A, Pappas G J. Stable flocking of mobile agents, Part I: Fixed topology [C]. In Proceedings of the 42nd IEEE International Conference on Decision and Control (IEEE Cat. No. 03CH37475), 2003, 2: 2010~2015.

[44] Olfati-Saber R. Flocking for multi-agent dynamic systems: Algorithms and theory [J]. IEEE Transactions on automatic control, 2006, 51 (3): 401~420.

[45] Ren W, Atkins E. Distributed multi-vehicle coordinated control via local information exchange [J]. International Journal of Robust and Nonlinear Control: IFAC-Affiliated Journal, 2007, 17 (10~11): 1002~1033.

[46] Desai J P, Kumar V, Ostrowski J P. Control of changes in formation for a team of mobile robots [C]. In Proceedings 1999 IEEE International Conference on Robotics and Automation (Cat. No. 99CH36288C), 1999, 2: 1556~1561.

[47] Beard R W, Lawton J, Hadaegh F Y. A coordination architecture for spacecraft formation control [J]. IEEE Transactions on control systems technology, 2001, 9 (6): 777~790.

[48] Ren W, Beard R W. Decentralized scheme for spacecraft formation flying via the virtual structure approach [J]. Journal of Guidance, Control, and Dynamics, 2004, 27 (1): 73~82.

[49] Bauso D, Giarré L, Pesenti R. Attitude alignment of a team of UAVs under decentralized information structure [C]. In Proceedings of 2003 IEEE Conference on Control Applications, 2003, 1: 486~491.

[50] Chen H F. Stochastic approximation and its applications [M]. Springer Science & Business Media, 2006.

[51] Fax J A, Murray R M. Information flow and cooperative control of vehicle formations [J]. IEEE

transactions on automatic control, 2004, 49 (9): 1465~1476.

[52] Ma C Q, Zhang J F. Necessary and sufficient conditions for consensusability of linear multi-agent systems [J]. IEEE Transactions on Automatic Control, 2010, 55 (5): 1263~1268.

[53] You K, Xie L. Coordination of discrete-time multi-agent systems via relative output feedback [J]. International Journal of Robust and Nonlinear Control, 2011, 21 (13): 1587~1605.

[54] You K, Xie L. Network topology and communication data rate for consensusability of discrete-time multi-agent systems [J]. IEEE Transactions on Automatic Control, 2011, 56 (10): 2262~2275.

[55] Meng Y, Li T, Zhang J F. Output feedback quantized observer-based synchronization of linear multi-agent systems over jointly connected topologies [J]. International Journal of Robust and Nonlinear Control, 2016, 26 (11): 2378~2400.

[56] Su Y, Huang J. Two consensus problems for discrete-time multi-agent systems with switching network topology [J]. Automatica, 2012, 48 (9): 1988~1997.

[57] Wang L Y, Yin G G, Zhang J F. Joint identification of plant rational models and noise distribution functions using binary-valued observations [J]. Automatica, 2006, 42 (4): 535~547.